Lecture Notes in Networks and Systems

Volume 340

The series "Lecture Notes in Networks and Systems" publishes the latest developments in Networks and Systems—quickly, informally and with high quality. Original research reported in proceedings and post-proceedings represents the core of LNNS.

Volumes published in LNNS embrace all aspects and subfields of, as well as new challenges in, Networks and Systems.

The series contains proceedings and edited volumes in systems and networks, spanning the areas of Cyber-Physical Systems, Autonomous Systems, Sensor Networks, Control Systems, Energy Systems, Automotive Systems, Biological Systems, Vehicular Networking and Connected Vehicles, Aerospace Systems, Automation, Manufacturing, Smart Grids, Nonlinear Systems, Power Systems, Robotics, Social Systems, Economic Systems and other. Of particular value to both the contributors and the readership are the short publication timeframe and the world-wide distribution and exposure which enable both a wide and rapid dissemination of research output.

The series covers the theory, applications, and perspectives on the state of the art and future developments relevant to systems and networks, decision making, control, complex processes and related areas, as embedded in the fields of interdisciplinary and applied sciences, engineering, computer science, physics, economics, social, and life sciences, as well as the paradigms and methodologies behind them.

Indexed by SCOPUS, INSPEC, WTI Frankfurt eG, zbMATH, SCImago.

All books published in the series are submitted for consideration in Web of Science.

More information about this series at https://link.springer.com/bookseries/15179

Rajiv Misra · Nishtha Kesswani ·
Muttukrishnan Rajarajan ·
Bharadwaj Veeravalli · Ashok Patel
Editors

Internet of Things and Connected Technologies

Conference Proceedings on 6th International Conference on Internet of Things and Connected Technologies (ICIoTCT), 2021

Springer

Editors
Rajiv Misra
Department of Computer Science
and Engineering
Indian Institute of Technology
Patna, Bihar, India

Muttukrishnan Rajarajan
City University of London
London, UK

Ashok Patel
Department of Computer Science
Florida Polytechnic University
Lakeland, FL, USA

Nishtha Kesswani
Department of Computer Science
Central University of Rajasthan
Jaipur, Rajasthan, India

Bharadwaj Veeravalli
Department of Electrical and Computer
Engineering
National University of Singapore
Singapore, Singapore

ISSN 2367-3370 ISSN 2367-3389 (electronic)
Lecture Notes in Networks and Systems
ISBN 978-3-030-94506-0 ISBN 978-3-030-94507-7 (eBook)
https://doi.org/10.1007/978-3-030-94507-7

This Springer imprint is published by the registered company Springer Nature Switzerland AG
The registered company address is: Gewerbestrasse 11, 6330 Cham, Switzerland

Preface

This book presents the proceedings of the Conference Proceedings on 6th International Conference on Internet of Things and Connected Technologies (ICIoTCT), 2021, that was held in collaboration with Indian Institute of Technology (BHU), Varanasi, India; California State University, San Bernardino, USA; Emlyon Business School, France, and International Association of Academicians (IAASSE), USA.

The ICIoTCT 2021 presented key ingredients for the fifth Generation Revolution. The ICIotCT 2021 conference provided a platform to discuss advances in Internet of Things (IoT) and connected technologies (various protocol, standards, etc.). The recent adoption of a variety of enabling wireless communication technologies such as RFID tags, BLE, ZigBee and embedded sensor and actuator nodes and various protocols such as CoAP, MQTT and DNS has made IoT to step out of its infancy. Now smart sensors can collaborate directly with machine without human involvement to automate decision making or to control a task. Smart technologies including green electronics, green radios, fuzzy neural approaches and intelligent signal processing techniques play important roles for the developments of the wearable healthcare systems. The ICIoTCT 2021 conference aimed at providing a forum to discuss the recent advances on enabling technologies and applications for IoT.

Due to the outbreak of COVID-19, this year's conference was organized as a fully virtual conference. This was an incredible opportunity to experiment with a conference format that we plan to continue in the future. In order to prepare ICIoTCT 2021, the organizing committees, reviewers, session chairs as well as all the authors and presenters have made a lot of efforts and contributions. Thank you very much for always being supportive to the conference. We could not have pulled off this convention without all of your hard work and dedication.

Organization

Program Chairs

Rajiv Misra	Department of Computer Science and Engineering, Indian Institute of Technology, Patna, Bihta-801106, Patna, Bihar, India
Veeravalli Bharadwaj	Department of Electrical and Computer Engineering, National University of Singapore, Block E4, Level 5, Room 42 4 Engineering Drive 3, Singapore-117583
Nishtha Kesswani	Department of Computer Science, CENTRAL UNIVERSITY OF RAJASTHAN, NH-8, Bandar Sindri, Dist-Ajmer, India
Ashok Patel	Department of Computer Science, Florida Polytechnic University, 4700, Research Way, Lakeland, FL 33805-8531, USA
Muttukrishnan Rajarajan	City University of London, Northampton Square, London, EC1V 0HB, UK

Contents

Machine Learning Based Adaptive Auto-scaling Policy for Resource Orchestration in Kubernetes Clusters

Abhishek Dixit(✉), Rohit Kumar Gupta, Ankur Dubey, and Rajiv Misra

Indian Institute of Technology Patna, Dayalpur Daulatpur, India
{adixit,1821cs16,ankur.cs17,rajivm}@iitp.ac.in

Abstract. The wide availability of computing devices has cleared the path for the development of a new generation of containerized apps that can run in a distributed cloud environment. Furthermore, the dynamic nature of workload demands elastic application deployment that can adapt to any scenario. One of the most popular existing container orchestration systems, Kubernetes, has a threshold-based scaling strategy that can be application-dependent and difficult to modify. Furthermore, its vertical scaling approach is disruptive, limiting deployment availability. The scaling decisions, instead of being proactive, are of reactive nature. In this work, our goal is to dynamically collect resource utilization of pods and predict future utilization for a period of time, and use the maximum utilization of that time window for proactive scaling, improving the overall resource utilization. We also contrast Kubernetes' built-in threshold-based scaling policy with a model-based reinforcement learning policy and the suggested LSTM Recurrent Neural Network-based prediction model. We demonstrate the benefits of data-driven rules, which can be combined with the Kubernetes container orchestrator.

1 Introduction

Container technology exploded in popularity with the debut of Docker in 2013. Docker is an essential component of the cloud ecosystem. Containerisation technology has swiftly become one of the hottest issues in the world of cloud computing due to its effective usage of computer resources and economic benefits. Shortly after the debut of Dockers in 2013, there was a flood of new container orchestrators aimed at reducing the complexity required in deploying, maintaining, and scaling containerised applications. One of these systems, the open source project Kubernetes, created by Google and now managed by the Cloud Native Computing Foundation (CNCF), has become the de facto standard for container management. Kubernetes, which offers container orchestration, deployment, and administration, is also crucial in cloud architecture. Because of its developer-centric container ecology features, Kubernetes has been the preferred option for container orchestration solutions as container technology has advanced. The main feature of Kubernetes is that it scales containerized applications up or down

R. Misra et al. (Eds.): ICIoTCT 2021, LNNS 340, pp. 1–16, 2022.
https://doi.org/10.1007/978-3-030-94507-7_1

based on the app's resource usage. The usage of resources changes based on the load demanded by the consumers. The load of the entire cluster is determined by the utilisation of each node. As a result, the advantages and disadvantages of the cluster's autoscaling method are critical [1–4].

2 Problem Statement

Threshold-based scaling strategies are application-dependent and difficult to fine-tune. The built-in autoscaler automatically suggests and modifies a pod's resource needs and limitations. Using the pods' usage data, it calculates a suggested value for the pod's resource demands and changes it to that number. The issue here is that if a pod is not running within the VPA's suggested range, it terminates the presently running version of the pod so that it may restart and go through the VPA admission procedure, which changes the CPU and memory demands for the pod before it can start. Because VPA only offers one suggested value for the time being, owing to the dynamic nature of load, one may wind up evicting pods too frequently, reducing availability and squandering many valuable resources. One of the reasons why a built-in vertical pod autoscaler is not utilised in production is because of this. We intend to forecast resource demands for application pods for a specific time period in the future. After analysing Kubernetes' general design and autoscaling approach, this predictive knowledge may be leveraged for vertical autoscaling that is more sustainable and reduces pod interruptions and waste of usable resources.

3 Theory and Related Work

3.1 Kubernetes Architecture

Kubernetes is made up of two parts: the master node and the worker node. The API Server, Kube-scheduler, and Controller manager are the three major fundamental components of the master node. The API Server component is in charge of replying to the user's management request, while the scheduler component connects the pod to the correct working node. The controller manager component, which consists of a group of controllers, is in charge of controlling and managing the corresponding resources. On the functioning node, two critical components, kubelet and kube-proxy, are operating. Kubelet is in charge of container life cycle management, as well as volume and network management. The Kube-proxy component is responsible for cluster-wide service discovery and load balancing [5,6].

The master node is in charge of the Kubernetes cluster's operation. It serves as the entry point for all administrative procedures. The master node hosts the control processes required by the whole system [7,8].

3.2 Kubernetes Threshold-Based Auto-scaling Policies

One of the most important characteristics of Kubernetes as a container orchestration platform is its ability to scale containerized workloads in response to changing conditions. This is known as auto-scaling. In a Kubernets cluster, there are three popular techniques for auto-scaling [9].

Horizontal Pod Autoscaler (HPA). Horizontal Pod Autoscaler controls the amount of pod replicas. Horizontal scaling is another name for it.

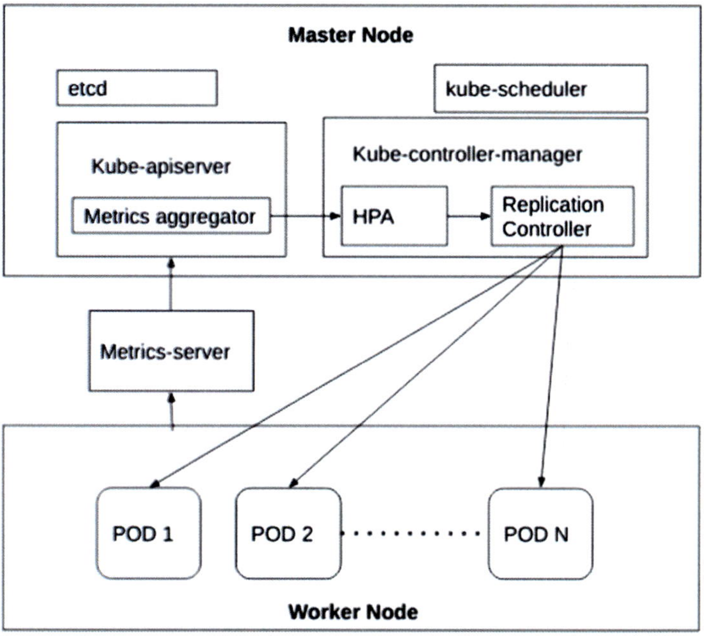

Fig. 1. Horizontal pod autoscaler (HPA)

The HPA controller collects utilisation information from the workload's pods' metrics-server and decides whether to change the number of copies of a running pod. In most situations, it does so by determining if adding or deleting a particular number of pod replicas will bring the current resource request value closer to the goal value. If there are n pods currently running and their individual CPU utilization is $U_1, U_2, ..., U_n$ then the average CPU utilization (U_{avg}) is the arithmetic mean of individual utilization.

$$U_{avg} = \sum_{i=1}^{n} \frac{Ui}{n} \tag{1}$$

If the target CPU utilization is U_{target} then the HPA controller adjusts the number of pod replicas n such that U_{target} and U_{avg} is as close as possible.

For example, consider a deployment that have a target CPU utilization $U_{target} = 60\%$ and number of pods $n = 4$, and the mean CPU utilization $U_{avg} = 90\%$. Let n' be the number of pods need to be added to make $U_{target} \approx U_{avg}$.

$$\implies \frac{U_{avg} * n}{n + n'} \approx U_{target}$$
$$\implies \frac{90 * 4}{4 + n'} \approx 60$$
$$\implies n' = 2$$

that means 2 more pod replicas need to be added.

Cluster Autoscaler. It works in the same way as the Horizontal Pod Autoscaler (HPA), but instead of adjusting the number of replicas of a pod in the cluster, it changes the number of worker nodes based on the load.

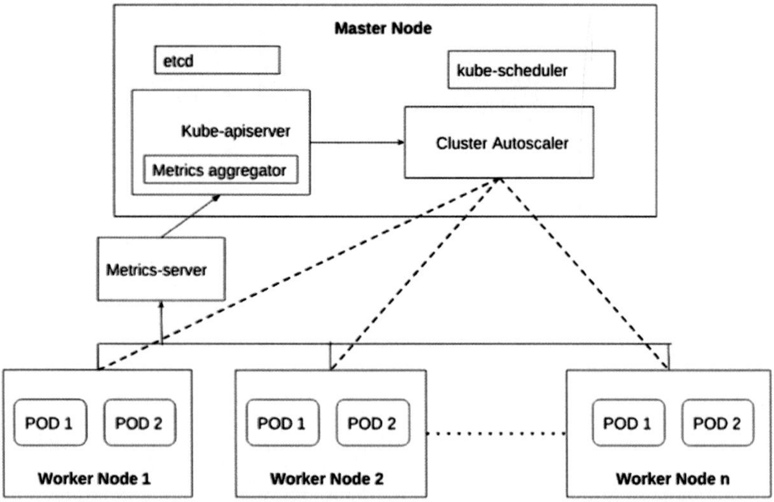

Fig. 2. Cluster autoscaler

The Cluster Autoscaler analyses the cluster to see whether there is a pod that cannot be scheduled on any of the available nodes owing to inadequate memory or CPU resources, or because the Node Affinity rule is in effect. Whether the Cluster Autoscaler discovers an unscheduled pod, it will examine its managed node pools to determine if adding any number of nodes can make this pod schedulable. If this is the case, it will add the necessary number of nodes to the pool if possible.

Vertical Pod Autoscaler (VPA). Vertical Pod Autoscaler (VPA) guarantees that the resources of a container are not under or overutilized. It suggests optimum CPU and memory requests/limits settings and may also automatically update them if in auto update mode, ensuring that cluster resources are utilised efficiently.

Vertical Pod Autoscaler is made up of three parts: the Recommender, which monitors pod resource utilisation using metrics from the metrics server and recommends optimal target values; the Updater, which terminates pods that need to be updated with newly predicted values; and the Admission Controller, which uses admission Webhook to assign the recommended values to newly created pods.

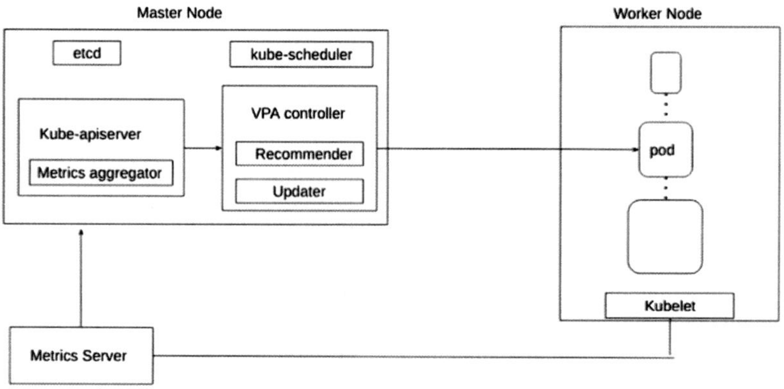

Fig. 3. Vertical pod autoscaler (VPA)

If a pod is specified with CPU request C_{req} and CPU limit C_{lim} and the recommended CPU request value is C_{newreq}, then the recommended CPU limit $C_{newlimit}$ is calculated proportionally:

$$C_{newlimit} = \frac{C_{lim}}{C_{req}} * C_{newreq} \tag{2}$$

For example, consider a pod p having $C_{req} = 50M$ and $C_{lim} = 200M$, then if the recommended request value $C_{newlimit} = 120M$, then:

$$C_{newlimit} = \frac{200}{50} * 120 = 480M$$

The main limitation of VPA is that A deployed pod's resource demands and limitations cannot be changed dynamically by the Kubernetes cluster. The VPA cannot add additional limitations to existing pods. It must evict pods that are not running within the anticipated range, and upon resuming, the VPA admission controller incorporates the suggested resource request and limit values into the specification of the newly formed pod.

3.3 Reinforcement Learning Scaling Policy

Many researchers have made significant contributions to enhancing the autoscaling policy [10,11] for various Kubernetes research topics. Fabiana Rossi [12] proposed a Reinforcement Learning based Kubernetes scaling policy to scale at run-time the number of conatinerized application pod instances. In Reinforcement Learning, an agent prefers activities that it found profitable in the past, which is referred to as exploitation. However, in order to uncover such rewarding behaviours, it must first investigate new activities, which is known as exploration. The RL agent determines the Deployment Controller state and updates the estimated long-term cost in the first phase, based on the received application and cluster-oriented metrics (i.e., Q-function) [13]. Furthermore, the Bellman equation replaces the simple weighted average of standard RL solutions (e.g., Q-learning) in updating the Q-function:

$$Q(s,a) = \sum_{s' \in S} p(s'|s,a)[c(s,a,s') + \gamma \min_{a' \in a} Q(s',a')] \tag{3}$$

Where γ is the discount factor c and p are cost function and transition probability respectively.

In the proposed autoscaler policy, Quality of Service requirements is expressed in terms of average response time with threshold $R_{max} = 80\,\text{ms}$.

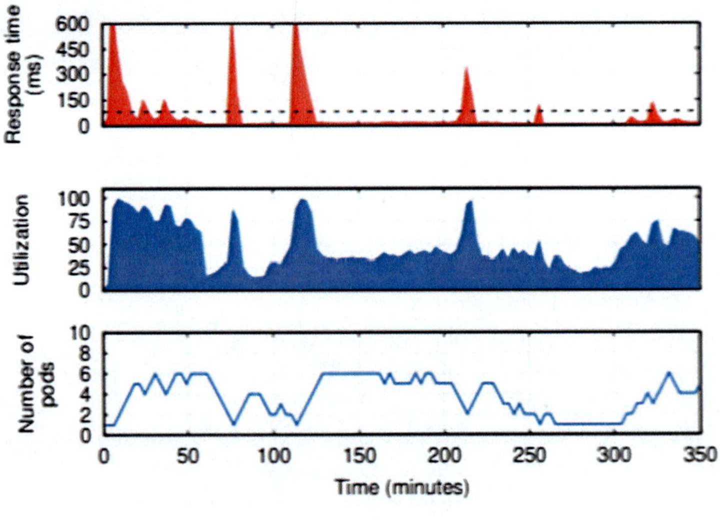

Fig. 4. Model-based RL autoscaler

4 Proposed Predictive Autoscaler

According to the circumstances described above, we need to dynamically gather resource consumption of pods and estimate future utilisation for a period of time, then scale based on the highest utilisation of that time frame.

We must gather historical resource consumption data and input it into our prediction module, which will offer resource recommendations for a period of time in the future. We shall then utilise the maximum resource request to avoid evicting pods too frequently.

The architecture of proposed predictive autoscaler is as follows:

Monitoring Module. Metrics Server is already present in an operational cluster. It is a scalable and efficient source for container resource measurements for Kubernetes' built-in autoscaling pipelines. These metrics can be used in our prediction model.

LSTM Prediction Module. The use of resources on each working node varies with time. It cannot reflect the real usage of node resources based just on node information at the current moment. As a result of examining the link between the changing trend of monitoring nodes and time, as well as using previous resource consumption data supplied by the monitoring module, a prediction model is created to anticipate resource usage for a period of time in the future.

Vertical Pod Autoscaling Module. Evict pods that are not running within the expected range, and when the VPA admission controller restarts, it incorporates the suggested resource request and limit values into the newly formed pod's specification.

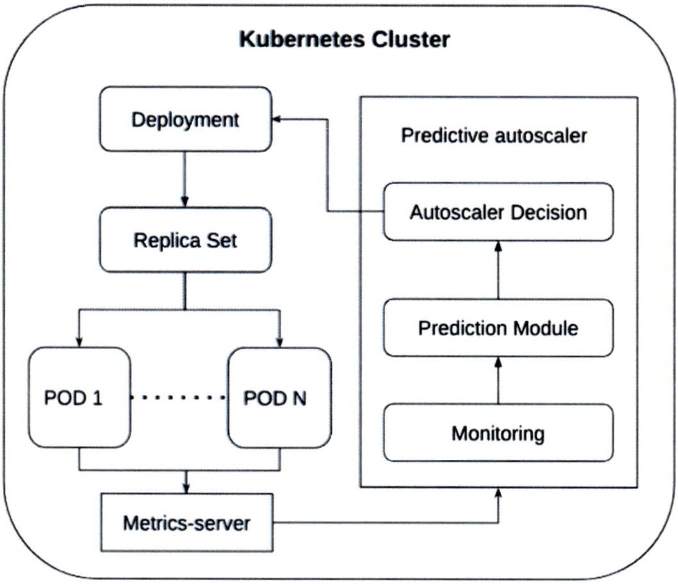

Fig. 5. Proposed predictive autoscaler architecture

5 Prediction Model

Long short-term memory (LSTM) [14] is a type of Recurrent Neural Network that has feedback that allows prior knowledge to be retained. When just short-term information is necessary to complete a task, traditional recurrent neural networks function effectively. Because of the vanishing/exploding gradients, RNN will struggle to represent a problem with long-term dependencies. The LSTM algorithm is a machine learning method designed to learn long-term dependencies. It keeps the information for a long time. In LSTM networks, memory blocks, rather than neurons, are connected across layers.

When compared to conventional neurons, an LSTM block contains distinct gates that allow it to store memory. A LSTM block comprises gates that determine the output and current state of the block. Each gate controls whether or not it is active by using the default sigmoid activation function units.

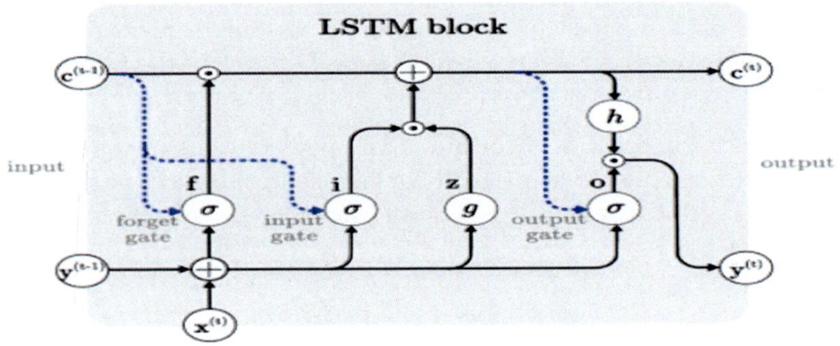

Fig. 6. LSTM architecture

There are three kinds of gates within a LSTM unit:

- **Forget Gate**: This gate determines information to be kept or thrown away from the block.
- **Input Gate**: This gate determines values from the input will be used update the memory state.
- **Output Gate**: This gate determines the next hidden state should be.

Each gate has default sigmoid function and a multiplication operation. Sigmoid function outputs to a range between 0 to 1 that is decided by the combination of $h(t-1)$ and $x(t)$. The output of the sigmoid function is then multiplied with the input to determine the gate's output. For example, if the output of sigmoid function is 1, then the gate's output will be equal to the input since input is multiplied by the result of sigmoid function. Each input vector to the unit is processes as following:

- The input vector $x(t)$ and the preceding hidden state vector $h(t-1)$ is chained to form another vector. The resultant vector will be used as the input in the three gates along with the tanh function.
- The forget gate uses the following equation to control what information to be kept within the unit:

$$f(t) = \text{sig}(W_f * [h(t-1), x(t)] + b_f) \tag{4}$$

where W is the weight and b is the bias.
- $\overline{C}(t)$, a new candidate values vector is calculated using:

$$\overline{C}(t) = \tanh(W_C * [h(t-1), x(t)] + b_C) \tag{5}$$

- The $\overline{C}(t)$ to be added to current cell is determined by multiplying it with $i(t)$ The equation of $i(t)$ is:

$$i(t) = \text{sig}(W_i * [h(t-1), x(t)] + b_i) \tag{6}$$

- The equation for the final current cell state $C(t)$ is:

$$C(t) = i(t) * \overline{C}(t) + f(t) * C(t-1) \tag{7}$$

- Output gate controls the amount the candidate value transferred into the next cell using:

$$o(t) = \text{sig}(W_o * [h(t-1), x(t)] + b_o) \tag{8}$$

- The equation for the final hidden state $h(t)$ is:

$$h(t) = \tanh(C(t)) * o(t) \tag{9}$$

6 Experimental Evaluation

Minikube, a tool for creating a local Kubernetes cluster, is being used for our experiment. For testing purposes, we installed a containerized Nginx web server on our Kubernetes cluster. We utilised Siege, a benchmarking programme, to generate traffic on our cluster, causing it to begin using cluster resources.

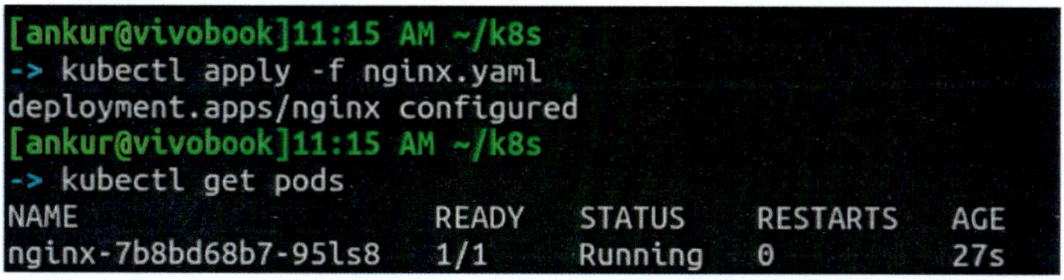

Fig. 7. Deploying nginx on minikube

Once the containerized app is deployed, we generated traffic using Siege tool as discussed above.

```
[ankur@vivobook]11:22 AM ~/k8s
-> minikube service nginx-service --url
http://192.168.99.100:31856
```

Fig. 8. Exposing app URL

```
[ankur@vivobook]11:26 AM ~/k8s
-> siege http://192.168.99.100:31856
New configuration template added to /home/ankur/.siege
Run siege -C to view the current settings in that file
** SIEGE 4.0.4
** Preparing 25 concurrent users for battle.
The server is now under siege...
```

Fig. 9. Generating traffic using siege

Once our cluster starts getting traffic, we can see in our Kubernetes dashboard that CPU and Memory utilization starts increasing for our pod. We collected these values at regular time intervals.

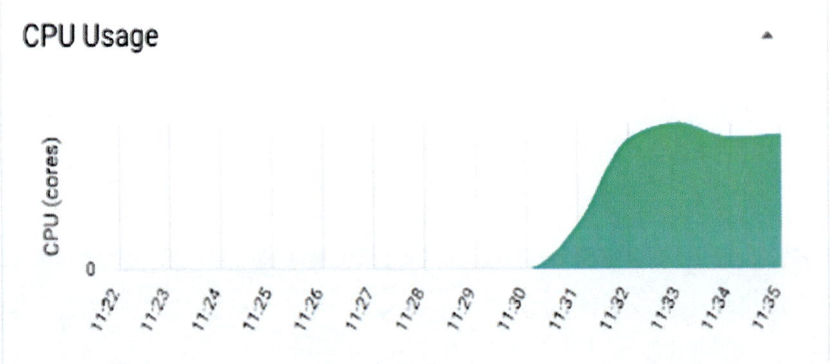

Fig. 10. CPU utilization after siege testing

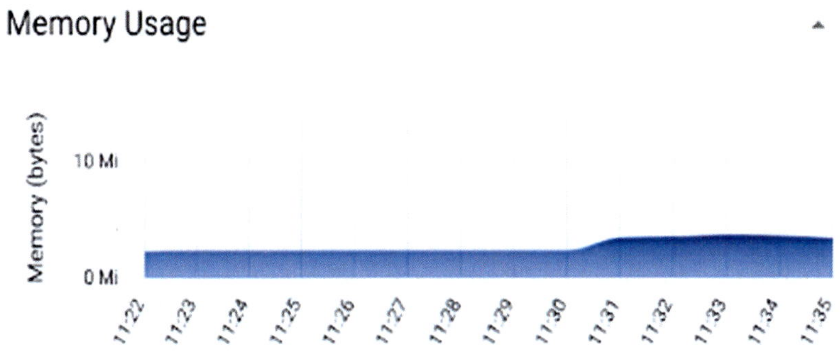

Fig. 11. Memory utilization after siege testing

6.1 Dataset

Real-time CPU and memory use data is required to develop the prediction model. The monitoring module, as previously mentioned, is used to create the dataset. We utilised Metrics Server, the most commonly used built-in open-source Kubernetes monitoring tool, which is already present in any Kubernetes cluster in production. We may send the measurements in whatever format needed and feed them into InfluxDB, an open-source time-series database. It receives metrics from Metrics Server and generates a time-series database with columns such as CPU Utilization percentage in Node, Number of pods in cluster, and so on, as well as the timestamp.

Plotting the complete CPU and Memory use dataset gathered using the aforementioned approaches, as well as its seasonal breakdown, for visualisation:

Fig. 12. CPU utilization

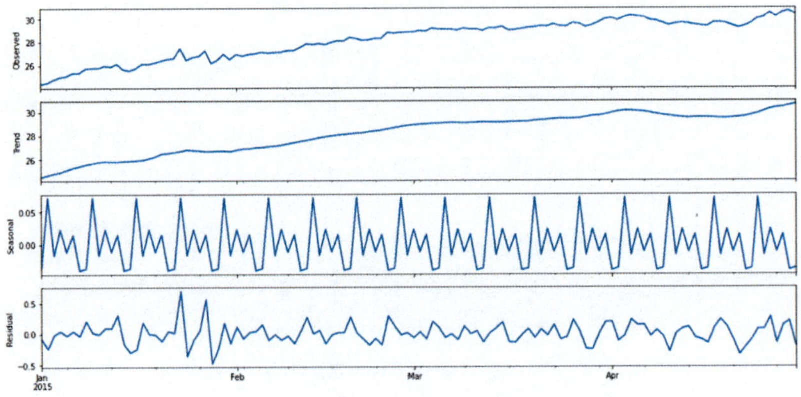

Fig. 13. CPU utilization components

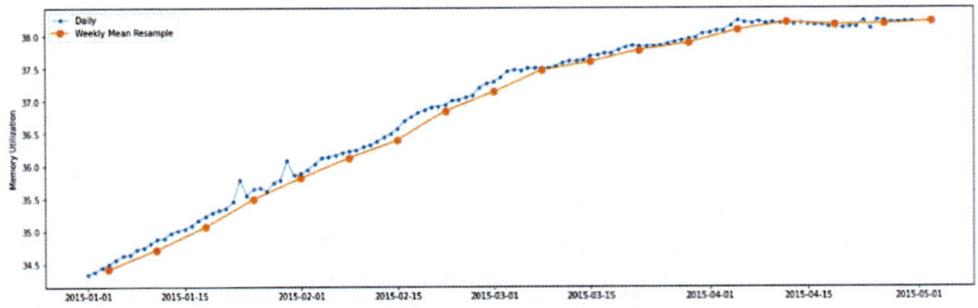

Fig. 14. Memory utilization

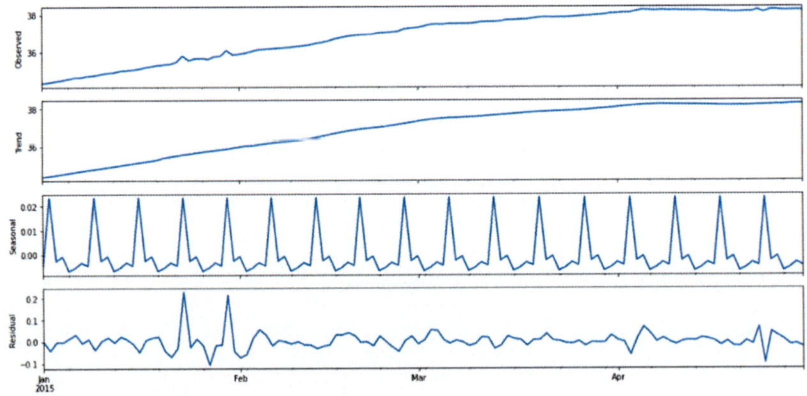

Fig. 15. Memory utilization components

The dataset has been split into training dataset that is 67% of the data and will be used for training the model, and the rest 33% data will be used for testing the model.

6.2 Evaluation Metric

We have used the Root Mean Squared deviation (RMSD) as a performance metric. The reason for choosing it over a strategy like Mean Absolute Error (MSE) because we do not desire to have large errors. RMSD uses the difference is squared terms; hence errors will reflect better in RMSD. RMSD is the difference between predicted and observed values first squared and then taken the mean of and lastly taken the square root of.

$$RMSD = \sqrt{\frac{\sum_{t=1}^{T}(x_{1,t} - x_{2,t})^2}{T}} \tag{10}$$

where $x_{1,t}$ and $x_{2,t}$ are actual observations series and estimated series respectively.

6.3 Prediction Results

We trained the LSTM RNN model with the collected utilisation data. With one visible input layer, one hidden layer and an output layer. The hidden layer has 4 LSTM blocks, and the prediction of output layer is a single value. The activation function used is the sigmoid function and Mean Squared Error was used as the loss function. After testing, the following results were obtained.

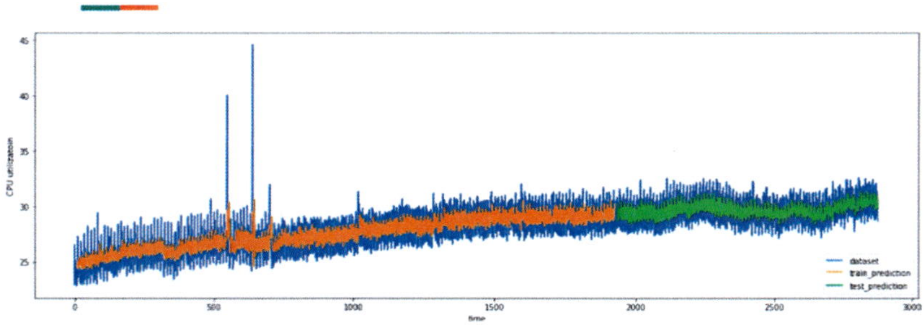

Fig. 16. CPU prediction

Our model performed well and the predictions are shown in the above diagram. Blue line represents the original data, green is the prediction on training data, and red is the prediction for test data. The RMSD error value of our CPU utilization forecasts is 1.49 for train data and 1.29 for test data, and for Memory utilization it is 0.29 for train data and 0.21 for test data.

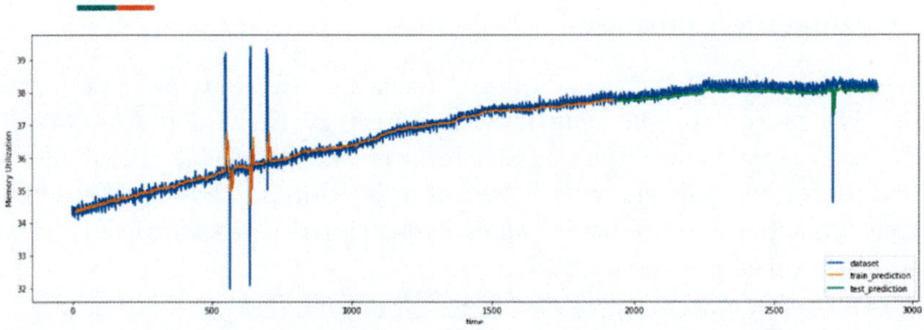

Fig. 17. Memory prediction

6.4 Comparison of Predictive Autoscaler with Default Autoscaler

The Kubernetes Vertical Pod Autoscaler uses statistical method for resource request recommendation. It does so by using weight buckets for historical resource usage and the weights diminishes exponentially. This is similar to the equation:

$$\hat{x}_{T+1 T} = \alpha x_T + \alpha(1-\alpha)x_{T-1} + \alpha(1-\alpha)^2 x_{T-2} + \cdots \quad (11)$$

with $\alpha = 0.5$

Using above prediction equation, the RMSD Error value our CPU utilization forecasts is 1.31, and for Memory utilization it is 0.27.

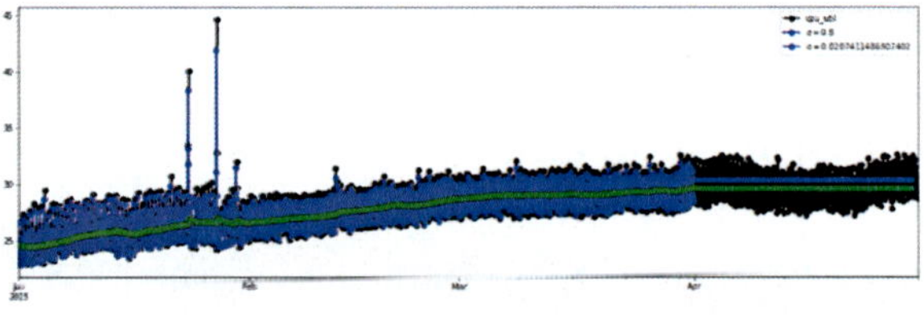

Fig. 18. CPU prediction using built-in VPA

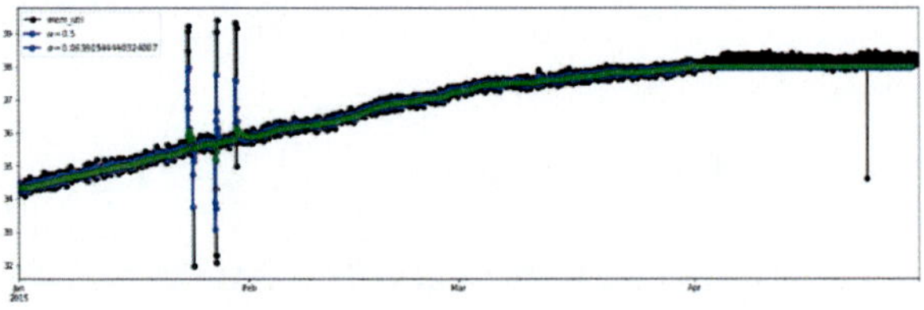

Fig. 19. CPU prediction using built-in VPA

As we can see in Fig. 16, 17, 18 and 19, the predictive autoscaler has performed better than the built-in autoscaler by slight margin. However, since predictive autoscaler uses data-driven approach, as it learns from more and more data, it's performance is expected to improve further. Also, the built-in VPA can only recommend memory values above 250Mi. So for apps that need lesser resource, our predictive model will ensure that resources are not getting under-utilized. The predictive LSTM Recurrent Neural Network model has no such lower or upper limit of prediction that makes it more flexible in real-life use cases.

Table 1. Comparison of predictive VPA with built-in VPA using RMSD as evaluation metric

Data	Predictive VPA (RMSD)	Built-in VPA (RMSD)
CPU utilization	1.29	1.31
Memory utilization	0.21	0.27

7 Conclusion

In this paper, we addressed autoscaling policy of the original Kubernetes cluster by analyzing the overall framework, aiming at the limitations of threshold-based scaling, vertical scaling and it's disruptive nature. A proactive LSTM Recurrent Neural Network based predictive autoscaler is proposed to optimize the prediction combined with the autoscaling strategy. However, this method may increase the computational burden due to its complexity. Considering the simple linear relationship between the increased computational burden and the number of nodes, it is completely acceptable. Optimized proactive autoscaling policy can improve the solution to a certain extent, and ultimately improve the resource utilization and Quality of Service (QoS) effectively. By collecting historical resource usage data of applications running on the Kubernetes platform, a combined prediction model is established to predict resource usage for a period of time in the future. This prediction data can be applied to the dynamic autoscaling module, so that we can improving the cluster resource utilization and Quality of Service.

References

1. Hu T, Yannian W (2021) A kubernetes autoscaler based on pod replicas prediction. In :2021 Asia-Pacific conference on communications technology and computer science (ACCTCS), IEEE
2. Imdoukh M, Ahmad I, Alfailakawi MG (2019) Machine learning-based auto-scaling for containerized applications. Neural Comput. Appl. 32(13):9745–9760. https://doi.org/10.1007/s00521-019-04507-z
3. Gupta RK, Bibhudatta S (2018) Security issues in software-defined networks. IUP J. Inf. Technol. 14(2):72–82
4. Ticherahine A, et al (2020) Time series forecasting of hourly water consumption with combinations of deterministic and learning models in the context of a tertiary building. In: 2020 international conference on decision aid sciences and application (DASA), IEEE

5. Gupta RK, Ranjan A, Moid MA, Misra R (2021) Deep-learning based mobile-traffic forecasting for resource utilization in 5G network slicing. In: Misra R, Kesswani N, Rajarajan M, Bharadwaj V, Patel A (eds) ICIoTCT 2020, vol 1382. AISC. Springer, Cham, pp 410–424. https://doi.org/10.1007/978-3-030-76736-5_38
6. Meng Y, Ra, R, Zhang X, Hong P (2016) CRUPA: a container resource utilization prediction algorithm for auto-scaling based on time series analysis. In: 2016 international conference on progress in informatics and computing (PIC), Shanghai, pp 468–472
7. Xie Y, et al (2020) Real-time prediction of docker container resource load based on a hybrid model of ARIMA and triple exponential smoothing. IEEE Trans. Cloud Comput. https://doi.org/10.1109/TCC.2020.2989631
8. Zhao H, Lim H, Hanif M, Lee C (2019) Predictive container auto-scaling for cloud-native applications. In: 2019 international conference on information and communication technology convergence (ICTC), Jeju Island, Korea (South), pp 1280–1282
9. Toka L, Dobreff G, Fodor B, Sonkoly B (2020) Adaptive AI-based auto-scaling for kubernetes. In: 2020 20th IEEE/ACM international symposium on cluster, cloud and internet computing (CCGRID), pp 599–608. https://doi.org/10.1109/CCGrid49817.2020.00-33
10. Gupta RK, Choubey A, Jain S, Greeshma RR, Misra R (2021) Machine learning based network slicing and resource allocation for electric vehicles (EVs). In: Misra R, Kesswani N, Rajarajan M, Bharadwaj V, Patel A (eds) ICIoTCT 2020, vol 1382. AISC. Springer, Cham, pp 333–347. https://doi.org/10.1007/978-3-030-76736-5_31
11. Gupta RK, Raji, M (2019) Machine learning-based slice allocation algorithms in 5G networks. In: 2019 international conference on advances in computing, communication and control (ICAC3), IEEE
12. Rossi F (2020) Auto-scaling Policies to Adapt the Application Deployment in Kubernetes
13. Fabiana R et al (2020) Geo-distributed efficient deployment of containers with Kubernetes. Comput. Commun. 159:161–174
14. Van Houdt G, Mosquera C, Nápoles G (2020) A review on the long short-term memory model. Artif. Intell. Rev. 53:5929–5955. https://doi.org/10.1007/s10462-020-09838-1

Transfer Learning Based Approach for Pneumonia Detection Using Customized VGG16 Deep Learning Model

Amit Ranjan(✉), Chandrashekhar Kumar, Rohit Kumar Gupta, and Rajiv Misra

Department of Computer Science and Engineering, Indian Institute of Technology Patna, Patna 801103, India
{amit_1921cs18,1811cs06,1821cs16,rajivm}@iitp.ac.in

Abstract. Pneumonia impacts seven percent of the world's population, contributing to 2 million children's deaths per year out of which India contributes about 23% of the global pneumonia burden, also the World Health Organization (WHO) estimates that pneumonia is responsible for one out of every three deaths in India. Pneumonia is normally detected by a highly-trained consultant's examination of a chest X-Ray radiograph. However, examining chest X-rays is a difficult task even for a professional radiologist. As a result, it is essential to create an automated method for detecting pneumonia to improve diagnosis accuracy. To address this, we develop a deep learning-based approach using a complex Visual Group Geometry VGG16 model to detect the presence of pneumonia. The proposed model has been trained on a collection of 5856 chest X-Ray images (4273 scans of pneumonia class and 1583 scans of normal class). For model evaluation, we have used accuracy, precision, recall, and f1 score metrics and the experiment performed on the public dataset shows that our proposed model achieves an accuracy of 98.28% with a minimized loss of 0.065, a precision value of 0.98, recall of 0.97, and F1 score of 0.976 respectively to detect pneumonia. Further, we analyze our model with different optimizers to observe any changes in the results. From the result, we can conclude that our developed model can also be used to diagnose pneumonia quickly with satisfactory performance accuracy.

Keywords: Deep learning · Pneumonia detection · VGG16 · Transfer learning

1 Introduction

Pneumonia is the most common cause of death for both children and adults all over the world. Elderly people over the age of 50, as well as children under the age of five, have a weakened immune system and are more vulnerable to disease. Pneumonia is the primary reason for death in kids under the age of 5, comprising about 16% of overall deaths [1] in this age range worldwide. According to the WHO, in a year, over four million worldwide prematurely deaths as a result of diseases. caused by domestic air contamination, such as pneumonia, and it affects about 7% of the world's population each year. India has the second-highest rate of pneumonia deaths in kids under the early ages, composes around

R. Misra et al. (Eds.): ICIoTCT 2021, LNNS 340, pp. 17–28, 2022.
https://doi.org/10.1007/978-3-030-94507-7_2

23% of the worldwide pneumonia load. As a result, early detection has become critical of such diseases in order to minimize the damage.

In medical applications, to increase the reliability and accuracy of diagnostic facilities, computer-aided diagnosis (CAD) technologies have been proposed and have been widely used throughout the last decade for pneumonia identification [4–6]. The main aim of CAD is to aid medical professionals in decision making and analysis by validating data using a computer system, which can help increase diagnostic accuracy, reduce specialist workload, and reduce reading variability. As a result, numerous computer algorithms for analyzing X-ray images have been proposed by researchers [7, 8]. A variety of computer-assisted diagnostic techniques [9, 10] have since been created to help in the processing of X-ray images. But these techniques are unable to offer enough knowledge to assist physicians in making decisions. In order to develop Machine Learning (ML) techniques, significant features are of paramount significance. As a result, most previous algorithms for designing CAD systems focused on image analysis relied on hand-crafted features [11, 12, 27].

Deep learning (DL) techniques surpassed standard machine learning methods in certain computer vision and medical imaging activities, such as target recognition and segmentation, such as X-ray image processing to analyze biological or abnormal structures of the patient's body [13, 14]. Deep learning models, especially convolutional neural networks (CNNs), had already shown promising results in object detection and segmentation, and as a result, they are commonly used in research [15]. The feature extraction process necessitates transfer learning approaches, in which pre-trained CNN models learn standardized features on large datasets like ImageNet and then transfer them to the appropriate task.

In this research, which was inspired by the accurate and most reliable efficiency of pneumonia detection using deep learning (DL), we propose a customized VGG-16 CNN model via an optimized layer architecture that can accurately identify pneumonia from chest X-Rays data. Further, in order to determine the perfect optimizer for interpreting X-Rays, a comparative analytical analysis of various optimizers was con-ducted. We used the accuracy, precision, recall, and f1 score metrics to evaluate the model. As seen in Fig. 1, the dataset used for this work has two classes.

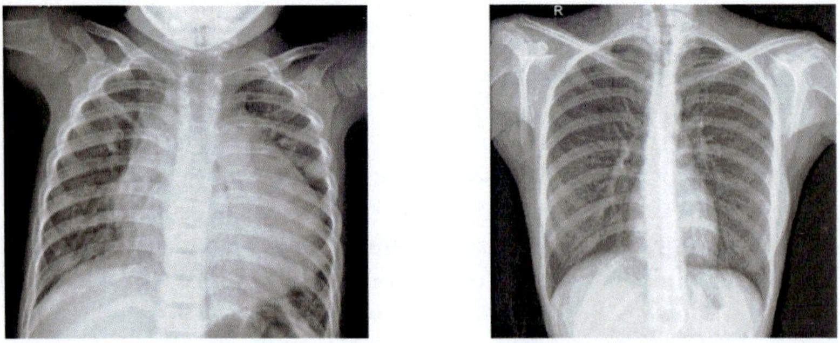

(a) Pneumonia (b) Normal

Fig. 1. Sample images of chest X-Ray with pneumonia (left) and normal (right) patients.

The remaining of the article is prepared in the following manner: the second section details some of the similar work. Explanation of our developed method is mentioned in Sect. 3. Section 4 presents and interprets some of the findings obtained utilizing our proposed models. The final section comprises the conclusion and future work.

1.1 Motivation

For pneumonia detection, one of the following procedures should be used: X-rays of the chest, a CT scan of the lungs, a chest ultrasound, and a chest MRI [2]. X-rays scans are now among the most effective tools in diagnosing pneumonia [3]. Identifying pneumonia in X-rays, on the other hand, is a difficult challenge that requires the presence of specialist radiologists. As a result, detecting pneumonia by interpreting chest X-ray can even be time-taking and inaccurate. The explanation for this is that a variety of other medical problems, such as lung cancer, extra blood, overlapping X-ray images of certain other diagnoses, and a variety of other benign anomalies, may all cause similar opacification in images. As a result, efficient X-ray reading is extremely desirable.

1.2 Problem Statement

We have a collection of x-ray images of patients and for each image we have to decide whether it belongs to Pneumonia or Normal patient. That means it is a two class (Pneumonia and Normal) classification where our aim is to detect pneumonia from X-Ray-image.

2 Related Work

A quick overview of several key contributions from the current literature is presented in this section. For enabling the usage of quantitative sequence data in electronic medical records, the authors [16] used traditional approaches (logistic regression, support vector machines (SVM), significant gradient boosting) and introduced techniques focused on multi-layer perceptron and recurrent neural networks (RNN). The results revealed that as opposed to many conventional machine learning techniques [29, 30], the deep learning-based model delivers the best results. Deep learning developments have currently proved to be useful in biological applications.

The authors of [17] suggested employing machine learning models to optimize the diagnosis and localization of pneumonia on X-rays images. Authors have used 2 CNNs including RetinaNet and Mask R-CNN in order to achieve their goal. The authors of [18] suggested a Deep Learning model for investigating lung cancer and pneumonia. The performance of their proposed enhanced AlexNet deep learning-based algorithm to identify X-Ray images in healthy and pneumonia classes using SVM was evaluated with certain other pre-trained models. Using a model trained on ImageNet, [19] used the principle of transfer learning in a deep learning environment to diagnose pneumonia. To localize pneumonia in Chest X-Rays, [20] introduced a Mask-RCNN (Region-based CNN), which was combined with image augmentation, which uses both global and local characteristics to segment pixels. The authors of [21] used to locate and spectrally identify pneumonia in Chest X-Rays using a gradient-based ROI localization method. In [22], an attention-guided mask estimation formula is proposed to detect important regions within the X-ray image that are characteristic of pneumonia. In their developed framework, the local features and global features of the network branches are combined to predict pneumonia. In [4], a gradient-based visualization approach is proposed to localize the region of interest (ROI) using heat maps in order to diagnose pneumonia. To estimate the disease risk, they used 121 densely connected layers. In [17], authors suggested a region-based CNN for lung image segmentation and image augmentation for pneumonia detection. In [23], authors utilize AlexNet model and GoogLeNet model for pneumonia detection. Centered on chest computed tomography (CT) X-Rays, authors in [24] suggested a new multi-scale heterogeneous 3D convolutional neural network. The authors in [25] suggested a hierarchical CNN framework for pneumonia diagnosis, using sine-loss function as loss function.

The best result mentioned from the above studies was 96.84% and 93.6%, respectively in identifying normal "versus" pneumonia cases using chest X-ray and deep learning techniques. As a result, there is plenty of space for improvement, either using new deep learning methods or the modification of current outsourcing models to increase classification accuracy.

3 Methodology

3.1 Dataset

A dataset of chest X-ray images collected from Kaggle [26] which is publicly accessible was used for training and testing of our model. In the original contains there were 5,856 scans of the human chest. Images have different dimensions originally. Each scan is labeled with one of two categories i.e. Normal class or Pneumonia class, based on the radiological records.

3.2 Data Preprocessing and Splitting

Original images present in the dataset vary in size and dimension. So, it needs to be established a base size for all images. We read the image one by one and resized it to 224 × 224, a fixed dimension for all images. We also passed each image to a preprocess_input function provided by the imagenet_utils library of Keras applications package, which smoothens our images to remove unwanted noise, if any. We collected all processed images in a list. We assigned the label ('0' for Normal and '1' for Pneumonia class) to each image and converted labels into one-hot encoding. We also shuffle the dataset using the shuffle function of the 'utils' library of keras's sklearn package, which ensures that each image participates independently in the model training without being affected by the image before it in the dataset. For training and testing our model, the dataset is splitted into train, validation, and test sets as mentioned in Table 1.

Table 1. Dataset description used for simulation.

Data	Normal	Pneumonia
Training samples	1223	3485
Validation samples	126	398
Testing samples	234	390
Total	1583	4273

3.3 Convolutional Neural Network and Transfer Learning

A convolutional neural network (CNN) is a multi-layer architecture that is a combination of convolution and pooling layers. These layers extract salient features from images and construct the best model to give the best results. Overall, a CNN can be described as image input passed to different types of layers and activation functions to learn a model, i.e. a fusion of extracting features and neural network classifiers meth-ods to learn a model as shown in Fig. 2.

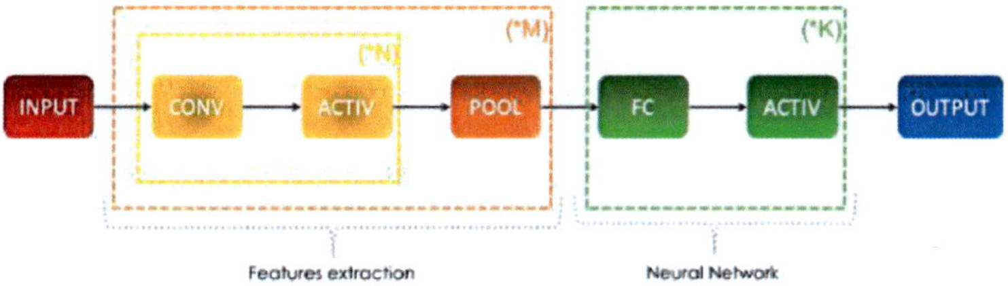

Fig. 2. Block diagram of CNN architecture

A Transfer learning [28] is a ML technique for reusing the starting point for a particular model on a different approach. The transfer learning starts by training the CNN model on the large base dataset using a base network structure. This output of the trained model is usually recognized as a pre-trained model. Then, the learned features from the first step are transferred to train on the second dataset. In this process, we use a pre-trained VGG16 model trained on a dataset similar to that of our current dataset so that we could extract high-level features from a large standard dataset. In this technique we remove the output layer of the pre-trained model, making some top layers non-trainable and adding some dense layers or adding some new convolutional layers over that to build our complex model. Our complex model learns low-level features from a new dataset and trains very accurately to give the best classification results.

3.4 Convolutional Neural Network and Transfer Learning

Simonyan and Zisserman [31] proposed the VGG16, a CNN-based model. It was among the most well-known deep learning models that participated in the 2014 ILSVRC exhibition. On the ImageNet dataset, the approach outperforms a top-5 accuracy rate of 92.7%.

There are 16 layers overall in the network [31]. Several 3 × 3 kernel-sized filters were adopted next to each other in VGG16, trying to replace huge kernel-sized filters being used in previous models. The neural network's depth is expanded by using several layers of kernels. This allows the network to recognize and interpret many complicated patterns from the data. VGG16 has 3 × 3 dimensional convolutional layers, 2 × 2 dimensional average-pooling layers, and a lastly fully connected layer in its architecture (Fig. 3).

Fig. 3. Layered architecture of our baseline VGG16 model.

We loaded the VGG16 model and removed the output layer of the model. We kept the first few layers trainable to learn high-level features of the image and froze the rest of the convolutional layers. Then we have added some fully connected layers to learn on our dataset. In the first two fully connected layers we added 1024 channels each and in the third, we added 512 channels. These layers with relu activation function can learn more complex functions and classify for better results. The final layer has softmax activation which finally gives output as probabilities for the two classes.

3.5 Convolutional Neural Network and Transfer Learning

In deep learning different types of optimization algorithms are available to compute loss. Some algorithms are described as follows:

3.5.1 SGD with Momentum

The simplest optimization algorithm is stochastic gradient descent with momentum [32], weight modification in batch SGD with movement according to the time step t.

$$W_{ij}^{l} = W_{ij}^{l} - \alpha \frac{\partial}{\partial W_{ij}^{l}} L(W, b) \tag{1}$$

$$b_{i}^{l} = b_{i}^{l} - \alpha \frac{\partial}{\partial b_{i}^{l}} L(W, b) \tag{2}$$

The problem with batch SGD with momentum is that the convergence scheme is very slow. To increase the convergence of optimization algorithm we used momentum but it is not beneficial as much as we want.

$$W_{ij,t+1}^{l} = W_{ij}^{l} - \alpha \frac{\partial}{\partial W_{ij,t}^{l} + \varepsilon} * \frac{\partial}{\partial W_{ij}^{l}} L(W, b) \tag{3}$$

The major advantage of using this optimization technique includes there is no need to tune the learning rate manually, the learning rate will change after every iteration and also the sparse and dense features taken care of by the adaptive learning rate.

3.5.2 Adam

It works well for convex as well as non-convex functions. It uses an exponential decay averaging for past squared gradients that we denoted by v_t and is similar to momentum; it keeps an exponentially decaying average of past gradients m_t:

$$m_t = \beta_1 m_{t-1} + (1 - \beta_1) * g_t v_t = \beta_2 v_{t-1} + (1 - \beta_2) * g_t^2 \tag{4}$$

$$\acute{m}_t = \frac{m_t}{1-\beta_1^t} \tilde{v} = \frac{v_t}{1-\beta_2^t} \tag{5}$$

Resulted from the above equation the update rule of Adam is shown in Eq. 6.

$$W_{t+1} = W_t - \frac{\alpha}{\sqrt{\tilde{v}-\varepsilon}} * \acute{m} \tag{6}$$

In Adam also, there is no need to tune the learning rate manually. The learning rate will change after every iteration. Sparse and Dense features take care of the adaptive learning rate. After performing various experiments on different optimization techniques we found that Adam is the best optimizing algorithm.

3.6 Evaluation Metrics

After the training process, our model was put to the test upon its test dataset. The accuracy, recall, precision, and F1 score were used to evaluate the results. This subsection goes into all of the evaluation metrics included in this work. When identifying normal and pneumonia cases, true positive (TP) represents the proportion of pneumonia X-rays predicted correctly as pneumonia, true negative (TN) represents the proportion normal X-rays correctly identified as normal, false positive (FP) represents the proportion of pneumonia X-rays wrongly identified as pneumonia, and false-negative (FN) represents the proportion of normal X-rays wrongly identified as normal.

3.6.1 Accuracy

It is calculated as the number of correct predictions over the total predictions. All predictions i.e. true or false predictions are used to calculate accuracy, precision, and recall.

$$Accuracy = \frac{TP + TN}{TP + TN + FP + FN} \tag{7}$$

3.6.2 Precision

It tells how many positive predictions are actually positive out of total positive predictions resulted from the model.

$$Precision = \frac{TP}{TP + FP} = \frac{TP}{Total\ Predictive\ Positive} \tag{8}$$

3.6.3 Recall

It tells how many positive predictions resulted by the model are actually positive out of total actual positive labels.

$$Recall = \frac{TP}{TP + FN} = \frac{TP}{Total\,Actual\,Positive} \tag{9}$$

3.6.4 F1 Score

This metric involves both precision and recall as shown below.

$$F1 = 2 * \frac{Precision * Recall}{Precision + Recall} \tag{10}$$

Generally, true negatives are not so important from an evaluation point of view. By elaborating on the above formula, we get that this metric prevents true negatives from affecting the score by giving less weight to it.

4 Experiments and Results

This section describes various tests that were implemented to build the best model for Pneumonia detection. We outline the tests and measurement measures being used to determine the model efficiency. The X-ray dataset [26] is used to demonstrate the tests and measurement methods used throughout the study. All the tests were carried out on a regular Linux PC with a 12 GB Nvidia GPU card, and the CUDA Toolkit.

The accuracy and loss curves of VGG16 with SGD optimizer are shown in Fig. 4. That is, the training data accuracy increases sharply from epoch 0 to epoch 2, reaching 74% for epoch 2, and the accuracy slope of test data reaches 75.9% for epoch 35.

For train results, a significant decline in the loss curve could be seen from epoch 0 to epoch 35, at which loss is close to 0.525. The loss curve of test results shows the same thing, with a loss of 0.508 for epoch 35.

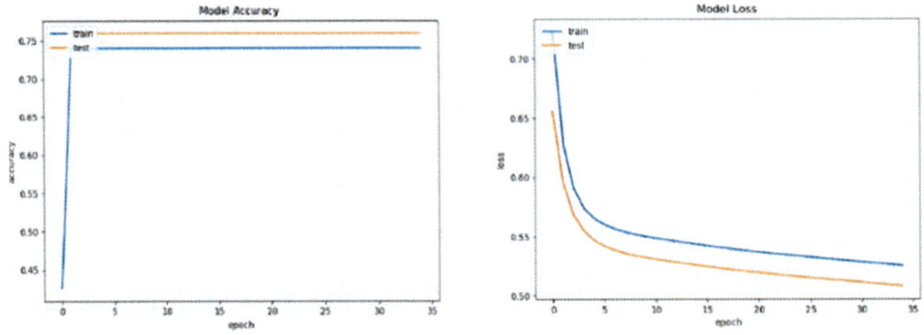

Fig. 4. Accuracy curve and loss curve of our proposed model using SGD optimizer.

With the Adagrad Optimizer, we see that training and testing accuracy converge at 93.7% approx. and losses converge at 0.16 approx. as shown in Fig. 5.

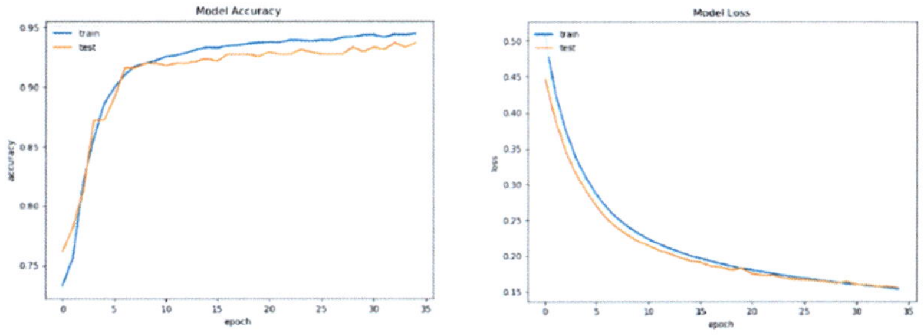

Fig. 5. Accuracy curve and loss curve of the proposed model using Adagrad optimizer.

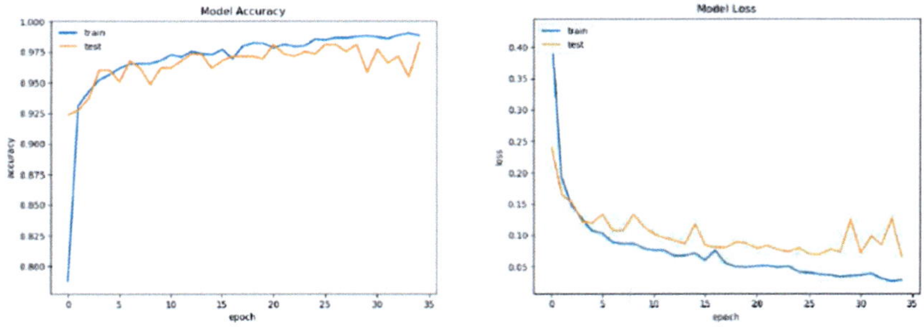

Fig. 6. Accuracy curve and loss curve of the proposed model using Adam optimizer.

With the Adam Optimizer, the accuracy graph of training data sharply increases from epoch 0 to epoch 2, and gradually increases until epoch 35, which is equivalent to 98.9%, and the accuracy graph of testing data is 98.2% for epoch 35 as shown in Fig. 6.

For train results, a significant decline in the loss curve is seen from epoch 0 to epoch 35, in which the loss is close to 0.029. For the loss of test results, in which the loss curve declines with minor fluctuation, epoch 35 equals 0.062. Now, let us consolidate training and testing results of different optimizers into a Table 2.

Table 2. Obtained training and testing results of different optimizers

Optimizer	Training accuracy	Testing accuracy	Training loss	Testing loss	Precision	Recall	F1 score
SGD	0.740	0.759	0.525	0.508	0.379	0.500	0.432
Adagrad	0.945	0.937	0.155	0.156	0.925	0.899	0.911
Adam	0.989	0.982	0.029	0.062	0.980	0.972	0.976

Finally, the obtained results show that the Adam optimizer is best for our model which got an accuracy score of 98.28% with a minimized loss of 0.065. For this optimizer, we got values of the precision, recall, and f1 score metrics for Normal and Pneumonia classes as shown in Table 3.

Table 3. Obtained precision, recall, and f1 score for normal and pneumonia classes

Class	Precision	Recall	F1 score
Normal	0.98	0.95	0.96
Pneumonia	0.99	0.99	0.99

From the above table, it can be observed that our model evaluates to be good for specific classes also when using adam optimizer that gives Precision value 0.980, Recall 0.972, and F1 Score 0.976 which are probably good for our Pneumonia Detection problem.

5 Conclusion and Future Work

Using deep CNN-based transfer learning methods, this study presents an automated diagnosis of pneumonia in X-ray scans. The studies were carried out using the X-Ray scan dataset [26] that comprises 5856 scans (4273 scans of pneumonia class and 1583 scans of normal class). Various scores, such as accuracy, recall, precision, and AUC ranking, were collected through experiments, demonstrating the effectiveness of our network model. The suggested framework was effective in achieving a classification accuracy of 98.28%, and precision, recall, and f1 score of normal images and pneumonia images with a value of 98%, 95%, and 96%; and 99%, 99%, and 99% respectively. In addition, hyper-parameter optimizations were considered and different optimization techniques were implemented, like stochastic gradient descent, Adagrad, and Adam optimizer to increase the efficiency of the model. The positive success of the customized VGG16 model trained on pneumonia detection shows that the model outperforms other optimizers as compared to Adam.

This research will be expanded in the future to detect and distinguish multi-class X-ray images. Additionally, using more advanced feature extraction methods based upon several recently developed deep learning models for biomedical image segmentation, the efficiency could be increased.

References

1. "World Pneumonia Day 2018." World Health Organization, World Health Organization, 12 November 2018. www.who.int/maternal_child_adolescent/child/world-pneumonia-day-2018/en/
2. Radiological Society of North America (RSNA) and American College of Radiology (ACR). "Pneumonia." Lung Inflammation - Diagnosis, Evaluation and Treatment. www.radiologyinfo.org/en/info.cfm?pg=pneumonia
3. World Health Organization (2001) Standardization of interpretation of chest radiographs for the diagnosis of pneumonia in children. No. WHO/V&B/01.35. World Health Organization
4. Rajpurkar P et al (2017) Chexnet: radiologist-level pneumonia detection on chest x-rays with deep learning. arXiv preprint arXiv:1711.05225 (2017)

5. Noor NM et al (2010) A discrimination method for the detection of pneumonia using chest radiograph. Computerized Med Imag Graph 34(2):160–166
6. Wang X et al (2017) Hospital-scale chest x-ray database and benchmarks on weakly-supervised classification and localization of common thorax diseases. IEEE CVPR
7. Avni U et al (2010) X-ray categorization and retrieval on the organ and pathology level, using patch-based visual words. IEEE Trans Med Imag 30(3):733–746
8. Pattrapisetwong P, Chiracharit W (2016) Automatic lung segmentation in chest radiographs using shadow filter and multilevel thresholding. In: 2016 International computer science and engineering conference (ICSEC). IEEE
9. Chen C-M et al (2013) Computer-aided detection and diagnosis in medical imaging
10. Qin C et al (2018) Computer-aided detection in chest radiography based on artificial intelligence: a survey. Biomed Eng Online 17(1):1–23
11. Poostchi M et al (2018) Image analysis and machine learning for detecting malaria. Transl Res 194:36–55
12. Das DK et al (2013) Machine learning approach for automated screening of malaria parasite using light microscopic images. Micron 45:97–106
13. Li J et al (2019) Study on the pathological and biomedical characteristics of spinal cord injury by confocal Raman microspectral imaging. Spectrochimica Acta Part A Mol Biomol Spectrosc 210:148–158
14. Winkel DJ et al (2019) Evaluation of an AI-based detection software for acute findings in abdominal computed tomography scans: toward an automated work list prioritization of routine CT examinations. Investig Radiol 54(1):55–59
15. Nijhawan R, Rishi M, Tiwari A, Dua R (2019) A novel deep learning framework approach for natural calamities detection. In: Fong S, Akashe S, Mahalle PN (eds) Information and Communication Technology for Competitive Strategies, vol 40. LNNS. Springer, Singapore, pp 561–569. https://doi.org/10.1007/978-981-13-0586-3_55
16. Ge Y et al (2019) Predicting post-stroke pneumonia using deep neural network approaches. Int J Med Inform 132:103986
17. Sirazitdinov I et al (2019) Deep neural network ensemble for pneumonia localization from a large-scale chest x-ray database. Comput Electr Eng 78:388–399
18. Bhandary A et al (2020) Deep-learning framework to detect lung abnormality–A study with chest X-Ray and lung CT scan images. Pattern Recogn Lett 129:271–278
19. Chouhan V et al (2020) A novel transfer learning based approach for pneumonia detection in chest X-ray images. Appl Sci 10(2):559
20. Jaiswal AK et al (2019) Identifying pneumonia in chest X-rays: a deep learning approach. Measurement 145:511–518
21. Wang X et al (2017) Chestx-ray8: hospital-scale chest x-ray database and benchmarks on weakly-supervised classification and localization of common thorax diseases. In: Proceedings of the IEEE conference on computer vision and pattern recognition
22. Guan Q et al (2018) Diagnose like a radiologist: attention guided convolutional neural network for thorax disease classification. arXiv preprint arXiv:1801.09927 (2018)
23. Lopez-Garnier S, Sheen P, Zimic M (2019) Automatic diagnostics of tuberculosis using convolutional neural networks analysis of MODS digital images. PloS One 14(2):e0212094
24. Xiao Z et al (2019) Multi-scale heterogeneous 3D CNN for false-positive reduction in pulmonary nodule detection, based on chest CT images. Appl Sci 9(16):3261
25. Xu S, Hao W, Bie R (2018) CXNet-m1: anomaly detection on chest X-rays with image-based deep learning. IEEE Access 7:4466–4477
26. Mooney P (2018) Chest X-Ray Images (Pneumonia). Kaggle, 24 March 2018. www.kaggle.com/paultimothymooney/chest-xray-pneumonia

27. Gupta RK, Choubey A, Jain S, Greeshma RR, Misra R (2021) Machine learning based network slicing and resource allocation for electric vehicles (EVs). In: Misra R, Kesswani N, Rajarajan M, Bharadwaj V, Patel A (eds) ICIoTCT 2020, vol 1382. AISC. Springer, Cham, pp 333–347. https://doi.org/10.1007/978-3-030-76736-5_31
28. Zhuang F et al (2020) A comprehensive survey on transfer learning. In: Proceedings of the IEEE 109(1):43–76
29. Gupta RK, Ranjan A, Moid M, Misra R (2021) Deep-Learning based mobile-traffic forecasting for resource utilization in 5G network slicing. In: Misra R, Kesswani N, Rajarajan M, Bharadwaj V, Patel A (eds) ICIoTCT 2020, vol 1382. AISC. Springer, Cham, pp 410–424. https://doi.org/10.1007/978-3-030-76736-5_38
30. Gupta RK, Misra R (2019) Machine learning-based slice allocation algorithms in 5G networks. In: 2019 International conference on advances in computing, communication and control (ICAC3). IEEE
31. Simonyan K, Zisserman A (2014) Very deep convolutional networks for large-scale image recognition. arXiv preprint arXiv:1409.1556 (2014)
32. Sutskever I et al (2013) On the importance of initialization and momentum in deep learning. In: International conference on machine learning. PMLR

Authenticate IoT Data for Health Care Applications Using ATSHA204 and Raspberry Pi

Navneet Kaur Brar(✉), Manu Bansal, and Alpana Agarwal

Electronics and Communication Engineering Department, Thapar Institute of Engineering and Technology, Patiala, Punjab, India
{navneet.kaur1,mbansal,alpana}@thapar.edu

Abstract. Internet of Things (IoT) has made life easier by connecting devices on one platform and communicates with other devices. However, the threat of malicious activities has also increased. The attackers intrude over the cloud malign the data or hack the device that proves a huge loss for the organization. Therefore, reliable systems are required to authenticate and secure the data in IoT. The paper presents a robust method to authenticate the data send from medical monitoring device. The Microchip ATSHA204 crypto IC is used to authenticate the data from its digital signature which locks the.csv data file generated by Keysight InfiniiVision MSO4024A Mixed Signal Oscilloscope. Raspberry pi 3B+ used as a controller in the technique. The coding is done in Python language.

Keywords: Internet of Things (IoT) · Secure hash algorithm (SHA) · Raspberry pi · Python · Oscilloscope

1 Introduction

In present scenario, Internet of Things (IoT) have collaborated numerous devices to communicate and exchange data over the internet. The data is being sent and received from the mobile phone, email or computer through internet to electronic devices such as phone, television, refrigerator or devices implant with temperature, pressure, passive infrared (PIR) sensors as shown in Fig. 1. IoT is widely used in everyday life in smart home appliances, medical devices, robotics, drones, medical and defense. The data used in IoT is stored over the cloud and has advantages such as resource availability, no requirement of specific software, economical and applications available on the internet [1]. Moreover, the data sent over the internet is confidential to the owner and great importance to the countries government as well. Many a time's rivals alter the data to affect good will of the company. Therefore, the authenticity, privacy and security of data are of prime importance.

Recently, researchers have proposed various techniques for the authenticity of the IoT data in various fields. For smart cards, data authenticity B.D. Deebak et al. used elliptic-curve lightweight cryptography to authenticate the IoT data and even from other vulnerabilities [2]. The IoT used in medical requires data protection for the patient's

R. Misra et al. (Eds.): ICIoTCT 2021, LNNS 340, pp. 29–38, 2022.
https://doi.org/10.1007/978-3-030-94507-7_3

privacy. M. Anuradha et al. used Advance Encryption Standard (AES) symmetric algorithm to protect the privacy of cancer patient's blood cells data [1]. Ali Shahidinejad et al. proposed a Light Edge based authentication protocol which constitutes IoT three layers: IoT device, Trust center and Service provider [3]. The protocol prevent attacks and has low communication cost [3]. MH Kashani gave a broader perspective of IoT in healthcare [13].

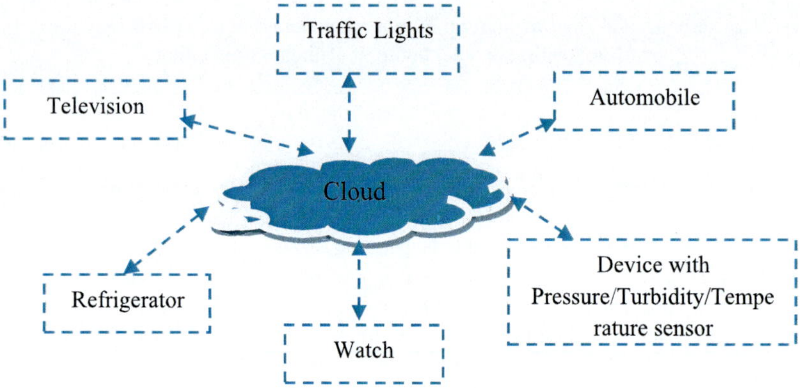

Fig. 1. Internet of Things (IoT)

In the following paper, a system is designed to authenticate the IoT oscilloscope data using ATSHA204 crypto IC for medical purposes. The Mixed Signal Oscilloscope (MSO) measures the patient heart rate, brain activity or any other pulse generated and stored in.csv file at regular intervals. The Raspberry pi 3B+ and ATSHA 204 attached to oscilloscope locks the.csv file with Digital signature generated by the crypto Authentic IC ATSHA204. Then the file is send to the hospital e-mail address only in case of emergency otherwise the record of each day is stored in.csv file and stored on the cloud for future reference with lock on it to protect the data. The data file can only be read by the receiver (hospital) only if they have key and setup to unlock the file. If there is an emergency i.e. drop in pulse or any negative result, message goes to the patient's family member and the doctor. This technique is beneficial for common people as there medical data is stored in a confidential manner. Moreover, the hospital and family is intimidated in case of emergency via mail and message respectively. Although the researchers such as FA Almalki [14], and SM Karunarathne [15] in 2021, P Huang [16] in 2019 and etcetera has used encryption to authenticate the healthcare data but none have used ATSHA204 IC system design to protect the healthcare data at hardware level. The main contribution of the propped technique is mentioned below:

- It is robust and economical to use for real time systems.
- It is portable and easy to install on any device.
- It consumes negligible power and less area.
- Software Tools used in the system are open source.

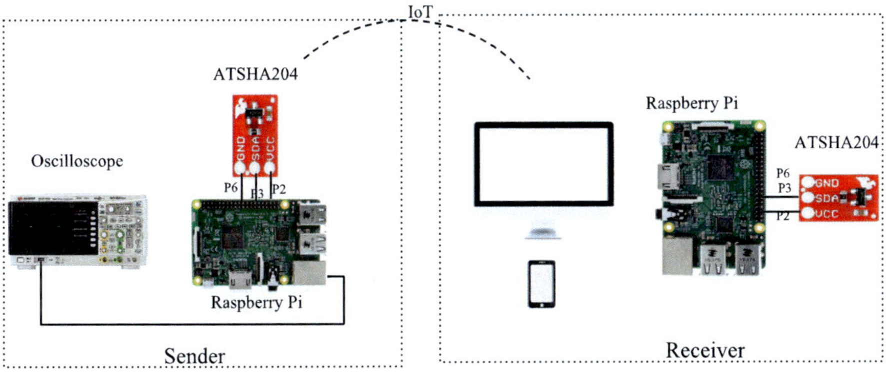

Fig. 2. Schematic of proposed system design

The remainder of the manuscript is as follows: Sect. 2 describes the hardware used in the system design. In Sect. 3 the proposed methodology is explained in detail. Section 4 demonstrates the experimental setup. Finally, the conclusion is drawn.

2 Hardware Used

To implement the proposed technique mainly Raspberry Pi 3B+ used as a controller; ATSHA204 IC used as an Authentication IC; MSO4024A Mixed Signal Oscilloscope to get the waveform; a Phone or PC is used to receive mail and message respectively. The Hardware used is explained in detail as follows:

2.1 ATSHA204

With the escalating attacks on devices security and privacy, organizations require methods to protect their devices. For which hardware security is always better and preferred but hardware solution being less economical, companies intend to adopt less effective software techniques. Therefore, Crypto Authentication IC ATSHA204 i.e. is cost effective, uses SHA-256 or HMAC algorithm to execute the challenge-response protocol [4]. ATSHA204 provides advance multilevel hardware security which comprises active shield over entire chip; memories inside chip are internally encrypted; supply tamper protection; internal clock and voltage regulation. As a result, the IC is widely used in industries for Brand Protection, Mange the license agreement and authorizes manufacturing at subcontractors.

The ATSHA204 IC uses I2C (Inter-Integrated circuits) communication protocol to transmit and receive challenge and response respectively through single pin. The physical parameters of IC are clock speed 1 MHz and input voltage range 2 to 5.5 V. ATSHA204 provides various applications such as accessory authentication in which fixed challenge response [4] is done, consumable authentication in which random challenge response is done, system anti-cloning in which anti-piracy is checked [4]. The ATSHA204 provides boot operation to the controller. The controller on startup sends the digital signature to the ATSHA204 and it respond to it by validating as authentic or not. Moreover, the ATSHA204 can be used for password checking in which CheckMac compares with internal password in ATSHA204 and generates a Boolean to determine if the password entered is correct or not [5].

2.2 Raspberry Pi

Raspberry Pi is a on board quad-core Arm Cortex with HDMI, Ethernet, USB port, Micro SD card slot, CSI camera connector, designed to be a low cost computer for educational purposes as shown in Fig. 3. In addition to it, it can be used for IoT applications, Robotics, Wireless server and many more [6]. The Pi support C, C++, Python, Java and other programming languages. It also supports various Linux machines. Raspberry Pi requires more power and current than Arduino or other microcontrollers which can be directly powered with computer USB. Therefore, we need a charger with 5 V, 1–2 A or can be given power through FTDI (USB to serial convertor). The rest of parameters can be studied from datasheet [6].

Fig. 3. Raspberry Pi 3 board

2.3 Oscilloscope

Oscilloscope is an electronic device to see a graphic display of voltage of any component or at any point [7]. The input to an oscilloscope is given through probes. There is one probe for each channel on the oscilloscope. It has a range of frequency up to which it can measure voltage a component. There are three types of oscilloscope based on type of signal it can measure: Analog Signal Oscilloscope, Digital signal oscilloscope, Mixed Signal Oscilloscope. One can measure different parameters of the voltage pulse such as amplitude, peak-to-peak value, root mean square (RMS) value, average value, sampling rate and etcetera depending on the model of oscilloscope we are using. Further features can be studies from user guide of the oscilloscope [8] (Fig. 4).

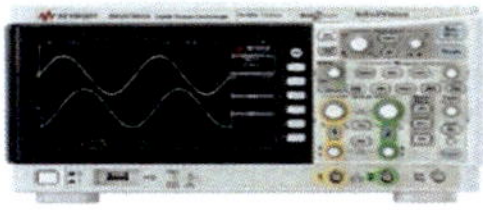

Fig. 4. Oscilloscope

3 Proposed Methodology

This system design could be divided into four steps: Configuration of Raspberry Pi and I2C in it; ATSHA204 personalization; Attach phone and computer with IoT to the setup. The Hardware connections are done as per shown in Fig. 2. The algorithm of the technique is as follows:

Algorithm: Authenticate IoT data using ATSHA204
Input: Pulse from the body to oscilloscope
Output: Mail/Phone to the doctor and family member
Method:
Step 1: In oscilloscope .csv file is generated
Step 2: ATSHA204 generates Digital signature
Step 3: Raspberry pi checks .csv data file:

If (Data > threshold limit) --- Implies Emergency

[Lower Threshold limit is 62 and Upper Threshold limit is 98 if heart beat is monitored, In case of ECG is upper limit is 198 and 122 is lower limit]

ATSHA and locks the data file with Digital Signature
/* Follows Step 4 */
Else
/* Step 1 and 2 are repeated */
End;

Step 4: Send the protected data file in email to hospital and message to family member for Emergency

The proposed system at block level is shown in Fig. 5 and the steps are explained in detail as follows:

Step I: *Raspberry Pi and I2C Configuration:* In order to configure the Raspberry Pi one would require a SD card Formatter, Etcher, Raspbian stretch. After this, in SD card Pi operating system is flash. The SD card is reinserted in the computer. Then to enable SSH in pi the ssh file is created without any extension. Another file, wpa_supplicant.conf file is created to setup the network wireless fidelity (WIFI) details as explained in [9]. Before supplying the power to the Pi, one should insert SD card in raspberry pi SD card slot. The ATSHA204 communicate with I2C protocol to Pi. After configuring the Raspberry Pi, I2C communication protocol is enable by following commands [10]:

Step 1: sudo apt-get install i2c-tools
Step 2: sudo apt-get install libi2c-dev
Step 3: cat /etc/modprobe.d/raspi-blacklist.conf
blacklist i2c-bcm 2708 --if this appears, follow step3.a; put # in front of this line
Step 3.a: sudo nano /etc/modprobe.d/raspi-blacklist.conf
Step 4: cat /etc/modules
snd-bcm 2835
i2c-bcm 2708 --add in the file
Step 4.a: sudo nano /etc/modules
i2c-dev -- add these two commands in the file
i2c-bcm 2708
Step 5: sudo reboot
Step 6: To check if I2c device is enable successfully
lsmod 1 grep i2c_ /* the list will come */

The presence of *i2c-bcm2708* will depict the i2c is enable [10]. In this manner Raspberry pi and I2C protocol is configured and set to use in system design.

Step II: *Personalize ATSHA204:* The Crypto IC is initialized and personalized with cryptonix hashlet (hashlet-1.1.0) library [11]. In the cryptotronix hashlet README.md file illustrates all the steps in detail. Firstly, root and configure of the IC is done. In root if those command not work for the system then try following commands.

- tar zxvf hashlet-1.1.0.tar.gz
- cd hashlet-1.1.0/
- sudo apt install gcc make libgcrypt11-dev
- sudo chmod a+x./configure -- to give execute permission to everyone
- ./configure
- Make

Then it is checked if the device is in which state: factory, personalize or tamper. The personalization is done which create a key file and store in ~/.hashlet [11]. After this the SHA204 operations random number generate, create mac (Message Authentication Code), hmac (hashlet MAC) and etcetera be performed.

Step III: *IoT in Computer and Phone:* To begin with, Pushbullet app is downloaded in the computer and mobile phone. The IoT is attached with Raspberry pi using python language. In computer, the data is send through mail from raspberry pi. The smtp lib for mail is installed in Pi terminal by following command:

- sudo apt-get install smtp mailutils
- sudo nano /etc./smtp/smtp.conf – Editor will open.

/* add the mail id of sender, password and receiver, TLS encryption is enabled.*/

Then in raspberry pi terminal sudo nano X.py code is written. The X should never be email as there are already email.py files in package that will create problem in the code [12]. For the phone, the pushbullet.sh file is created and the API token key is added from pushbullet app. Then, the code is written in python in nano editor. Make sure the raspberry pi time and date is up-to-date; if not can done by command "sudo date -s "29 May 2019 11:58:00".

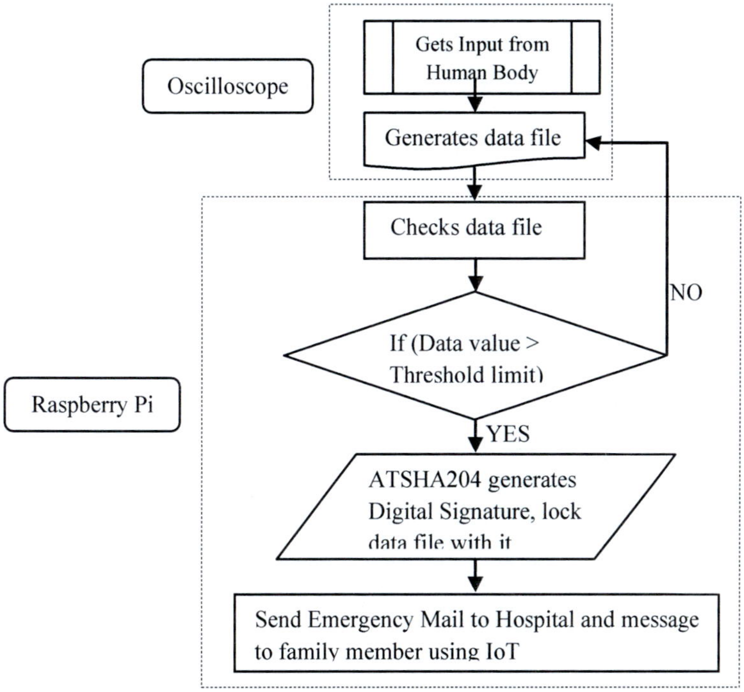

Fig. 5. Flow diagram of proposed technique using IoT

Therefore, in the Sect. 3 the methodology is described in particular. These are the steps to be followed to implement the authentication of IoT data with oscilloscope for healthcare applications.

4 Experimental Setup

To implement the proposed system design one would require putty software to use raspberry pi terminal in a computer. The Python 2.7 version installed in raspberry pi terminal that is accessed from putty in ssh mode. In hardware Keysight MSO 4024A, Raspberry Pi 3B+, Microchip ATSHA204, phone and a computer is needed. The IoT coding is done in python language with installed smtplib library in it. The ATSHA204 works with cryptotronix hashlet github library.

After hardware connections, at input side the ATSHA204 and Pi is connected to oscilloscope. The IoT code is already dumped in raspberry pi. The oscilloscope generates.csv data file and is sent to the computer through Ethernet or taken by pen drive. The setup at input side is shown in Fig. 6.

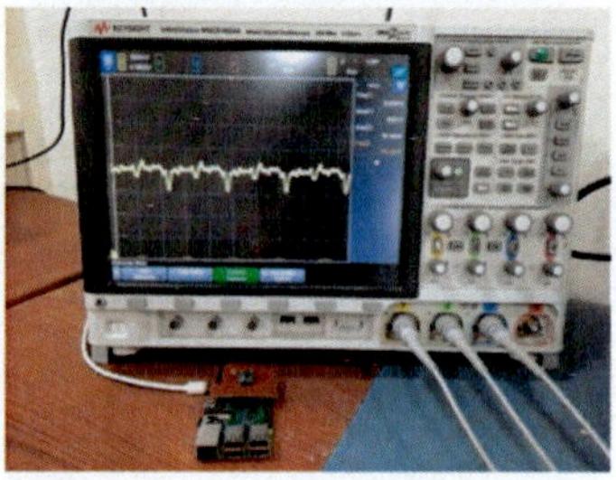

Fig. 6. System at sender side

At the output side, a computer, phone, Pi and ATSHA204 is there. The data file is opened by entering key in ATSHA204 that authenticate it and opens the file. The entire setup at receiver side is shown in Fig. 7.

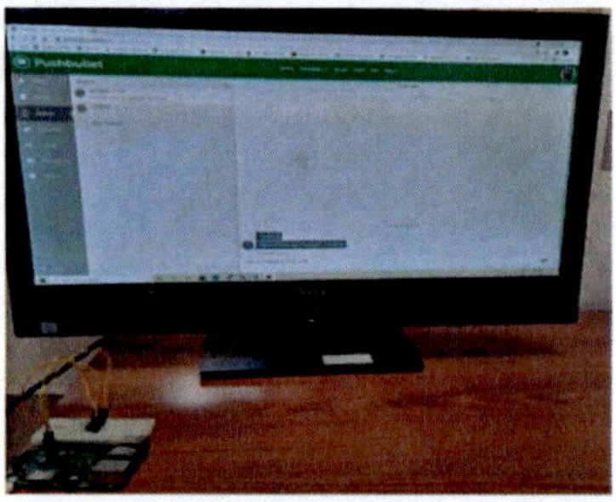

Fig. 7. System design at receiver side

The system performance is measured in terms of memory, area and cost. The memory is the raspberry pi accessed through putty. The cost and area of the system includes two raspberry pi and ATSHA204, as the oscilloscope, computer and phone are already used by the user. The parameters are shown in Table 1.

Table 1. Performance parameters

Constraints	Value
Overall cost	Rs 5600/- (per system)
Area	85.7 × 56 mm
Memory	100 MB

5 Conclusion

When security and privacy of data, in particular the IoT data communicated over the internet, has become matter of concern. This paper demonstrates an authentication technique to protect the health care data of the patients, which is sent to hospital and family member using IoT. The authentication is done using ATSHA204 unique Digital Signature. The data file is locked by this digital signature. It can only be opened at receiver side if the receiver has the key and system setup, which makes it confidential while communicating over the internet. However, sharing the key is usually challenging due to threat of key leakage. But in this scenario need of specific system design provides extra protection to the data which will be accessible only to the authorized person. Therefore, this methodology prevents data from malicious activities.

Acknowledgement. The authors are grateful to Ministry of Electronics and Information Technology (MeitY) for the financial support through 'SMDP Chip to System Design' Project. The authors would also like to thank Director, Thapar Institute of Engineering and Technology, Patiala, Punjab, for providing necessary resources to carry out this research.

References

1. Anuradha M et al (2021) IoT enabled cancer prediction system to enhance the authentication and security using cloud computing. Microprocess. Microsyst. 80:103301
2. Deebak BD, Fadi AT (2021) Lightweight authentication for IoT/Cloud-based forensics in intelligent data computing. Futur Gener Comput Syst 116:406–425
3. Shahidinejad A, Ghobaei-Arani M, Souri A, Shojafar M, Kumari S (2021) Light-edge: a lightweight authentication protocol for IoT devices in an edge-cloud environment. IEEE Consum Electron Mag 1
4. http://ww1.microchip.com/downloads/en/devicedoc/Atmel-8740-CryptoAuth-ATSHA204-Datasheet.pdf
5. http://ww1.microchip.com/downloads/en/AppNotes/Atmel-8794-CryptoAuth-ATSHA204-Product-Uses-Application-Note.pdf
6. https://www.raspberrypi.org/documentation/hardware/computemodule/datasheets/rpi_DATA_CM3plus_1p0.pdf
7. https://en.wikipedia.org/wiki/Oscilloscope
8. http://literature.cdn.keysight.com/litweb/pdf/54709-97048.pdf?id=2265134
9. https://www.raspberrypi.org/documentation/configuration/
10. https://radiostud.io/howto-i2c-communication-rpi/

11. https://github.com/cryptotronix/hashlet
12. https://bc-robotics.com/tutorials/sending-email-using-python-raspberry-pi/
13. Kashani MH, Madanipour M, Nikravan M, Asghari P, Mahdipour E (2021) A systematic review of IoT in healthcare: applications, techniques, and trends. J Network Comput Appl 192:103164
14. Almalki FA, Soufiene BO (2021) EPPDA: an efficient and privacy-preserving data aggregation scheme with authentication and authorization for IoT-based healthcare applications. Wirel Commun Mob Comput 2021
15. Karunarathne SM, Saxena N, Khan MK (2021) Security and privacy in IoT smart healthcare. IEEE Internet Comput 25(4):37–48
16. Huang P, Guo L, Li M, Fang Y (2019) Practical privacy-preserving ECG-based authentication for IoT-based healthcare. IEEE Internet Things J 6(5):9200–9210

Randomised Key Selection and Encryption of Plaintext Using Large Primes

Soumen Das[1(✉)], Souvik Bhattacharyya[2], and Debasree Sarkar[1]

[1] Indian Institute of Technology Kharagpur, Kharagpur, WB, India
[2] The University of Burdwan, Golapbag, Burdwan, WB, India

Abstract. Study of ciphertext plays a major role in Cryptanalysis. Functionally correct realization of a cryptosystem does not ensure the confidentiality of message always. Most classical cryptosystem are not quantum secure. Shor algorithm is the only polynomial time factorization attack which can break RSA in very less computation time. On the other hand side channel attack and fault attack is sufficient for information leakage. In this type of attack, the attacker neither uses the cipher text nor does the plain text rather consider the power trace, amount of memory used by the cryptosystem, timing variations for different cryptographic operations etc. So it is very important to make the cipher different in every run which makes it non deterministic hence the power analysis should be different in every run. The main idea behind this paper is to introduce randomized key to encrypt the plain text one character at a time.

Keywords: Dynamic key generation · Randomized key selection · Brute force attack · Factorization attack · Side channel attack · LUT

1 Introduction

Any classical cryptosystem is based on three criteria Encryption algorithm, Decryption algorithm and most importantly key generation. RSA [1–3, 10] is one of the most trusted public key cryptosystem for very large size key (commonly 1024 bits/2048 bits etc.). In RSA (n = p * q (product of two primes p and q), e (encryption key)) are public whereas d (decryption key) is private. Creation of cipher and retrieval of plaintext both are based on modular exponentiation.so factorization attack [5, 9, 13, 21, 22] used the multiplicative property between e and d as e * d = 1 mod phi_n, where phi_n = (p−1) * (q−1) [product of two non primes] such that d can easily be derived from d = (k * phi_n + 1) /e. Now the idea is how to hide d or produce different e to generate different d in different run which produces different cipher on same message each time. So non

R. Misra et al. (Eds.): ICIoTCT 2021, LNNS 340, pp. 39–46, 2022.
https://doi.org/10.1007/978-3-030-94507-7_4

deterministic behavior of cipher resist side channel attack [6, 7] (difficult to trace of power spectrum) fault attack [17, 18, 23] and brute force attack (probably take very very long time because of the use of exponential time key finding algorithm).

2 Proposed Methodology

In our literature we introduced dynamic key generation procedure by using two very large prime p and q where e is co-prime to Phi_n (GCD (e, phi_n) = 1). So we have generated list of possible candidates for encryption key e and select one of them randomly to encrypt the plain text characters one at a time. But the difficulty is the characters are string alphabets, so encryption is not possible directly. As a solution we convert the plain text character into their corresponding ASCII values one at a time then perform encryption operation by using randomized key selection from the set of possible candidates to introduce randomness and non-deterministic behavior of cipher. The no of possible keys (cardinality) depends not only on the prime numbers (p, q) also on the different runs over same plain text message. As for example if we take p and q as two 5 bit numbers, randomly two values assigned [(p = 29, q = 23) for 1st iteration with 117 possible keys, (p = 67, q = 71) for 2nd iteration with 223 possible keys] and so on. One of the key is selected randomly from list of possible keys to encrypt the characters of plain text one at a time. As a result no one can predict what is the key on different run regardless of same input provided every time which leads to the creation of different ciphers on every run with same plaintext input. In real time implementation size of prime numbers (p, q) never be 5 bits value, generally taking at least 200 bits (say 256 bits/512bits/1024 bits/2056 bits). That means the randomness of key selection increases as the input bit size of prime number increases. In other words cardinality of key array (No of possible keys for random selection) directly proportional to cardinality of input prime numbers (size of p and size of q).Cardinality of the key array strictly less than half of the cardinality of the product of two primes. Most interestingly we have to notice that on different run changes of the cardinality of key array, for same cardinality of prime inputs which makes the cipher non-deterministic in nature (Figs. 1, 2, 3, 4 and Table 1).

Proposed Algorithm:

Input: Plain text, Large Prime Numbers (bit size)
Output: Encrypted text

1. Generate Large prime p of>200 bits (256 bit/ 512 bit /1024 bit/ 2048 bit)
2. Generate Large prime q of>200 bits (256 bit/ 512 bit /1024 bit/ 2048 bit)

3. Calculate
   3.1 $n = p \times q$
   3.2 phi_n= $(p-1) \times (q-1)$ or $(p-1) \times (q+1)$ or $(p+1) \times (q-1)$ or $(p+1) \times (q+1)$ //product of non-primes

4. Take Plain text as input string as commandline arguments
   4.1. Plain text=[ ] // Input string
   4.2. Convert each character of Plain text to its corresponding ASCII value.
   4.3. Store all ASCII values inside a List ASCII = [ ] // initially empty

5. Possible set for key generation (for encryption)
   E_KEY = [ ] // initially empty
      for k in range (2,phi_n//2):
         e_k = Generate- key(GCD(k,phi_n) = = 1;such that ∄ any k which divides phi_n)
         E_KEY =E_key.append(e_k)

6. Key selection form E_KEY set and Encryption one character at a time
   Cipher = [ ] //initially empty
   LUT =[ ] // initially empty
   for j in Plain text :
      temp = pick (plain text [j])
      e_{key} = Random-select ($e_j \in$ E_KEY [ ])
      LUT= LUT.append (e_{key}) //{e_1, e_2 , e_3 ,..... e_n }
      cipher_text = temp $^{e_{key}}$ mod n
      cipher_text = convert to string (cipher_text) // ASCII to String conversion
      Cipher = Cipher.append (cipher_text)

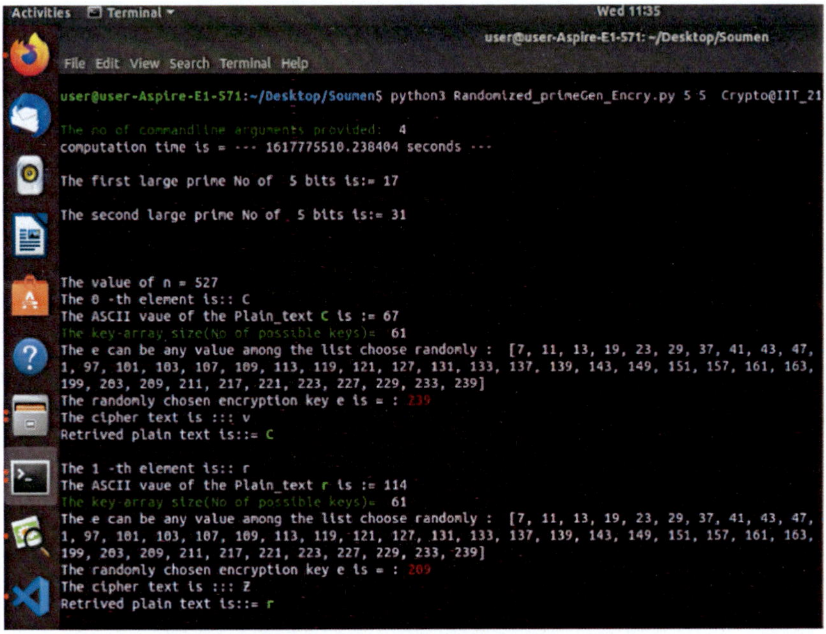

Fig. 1. Experiment output (when p = 5 bits, Q = 5 bits)

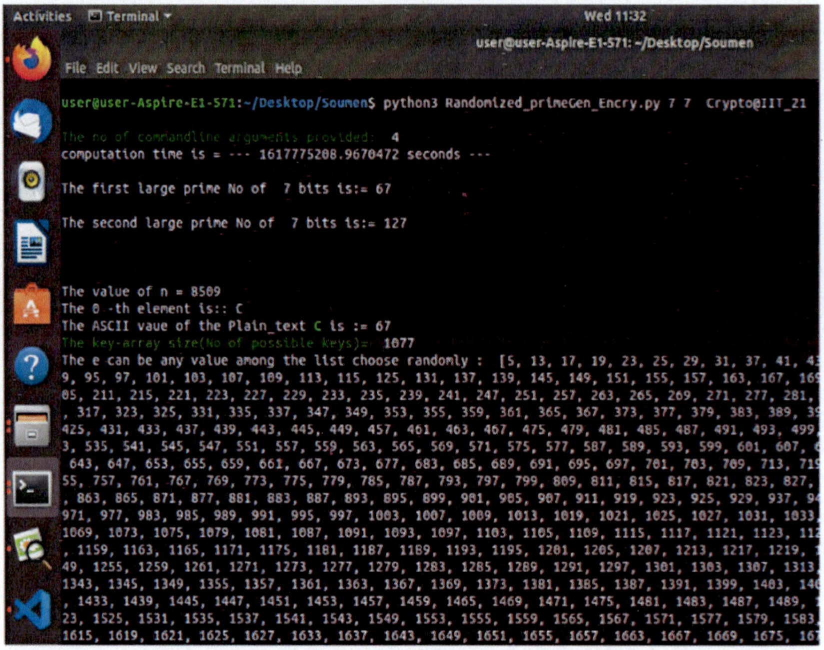

Fig. 2. Experiment output (when p = 7 bits, Q = 7 bits)

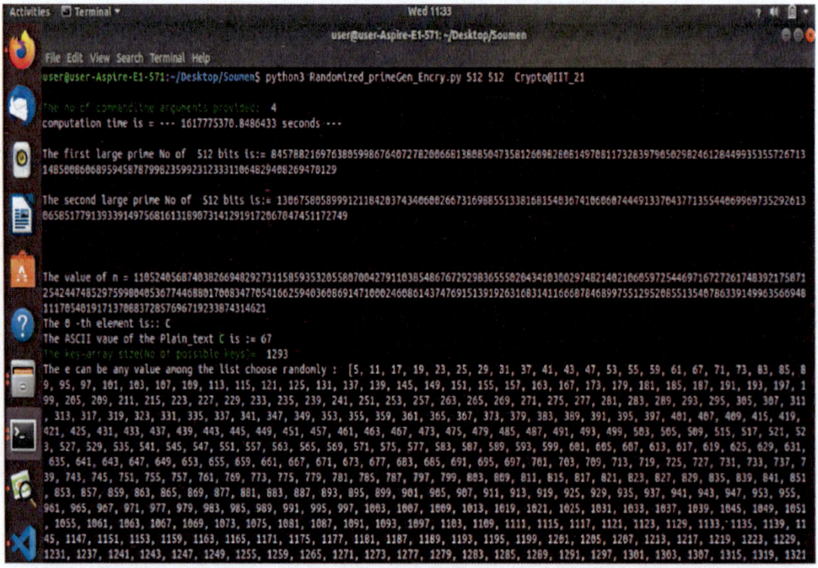

Fig. 3. Experiment output (when p = 512 bits, Q = 512 bits)

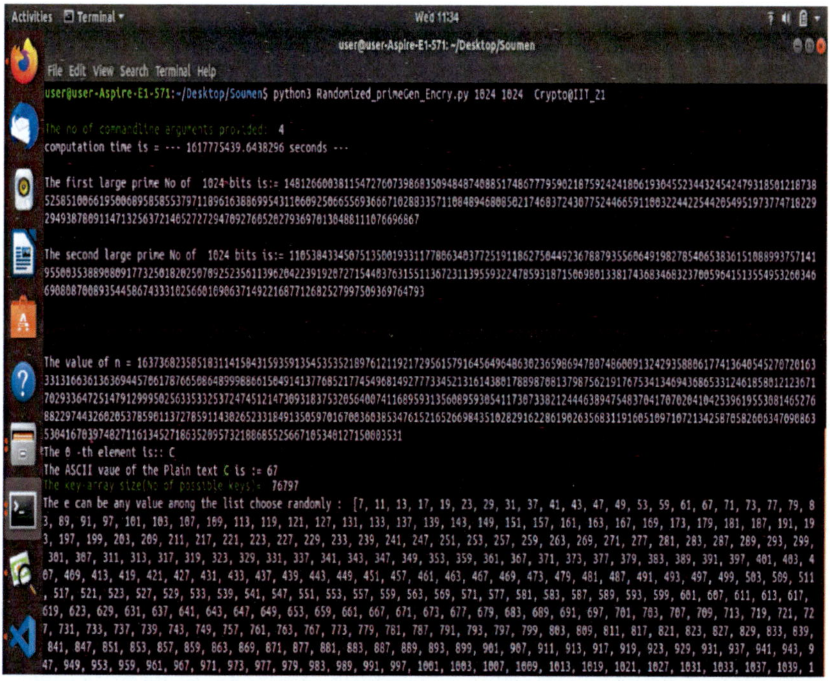

Fig. 4. Experiment output (when p = 1024 bits, Q = 1024 bits)

Table 1. Comparison table for key generation

No of bits for 1st prime (P)	No of bits for 2nd prime (Q)	Value of P	Value of Q	Product of P and Q n =P*Q	No of Possible keys = |KEY [ ]|
5	5	29	23	667	117
....	....	....	....	....	....
7	7	67	71	4757	447
7	7	97	89	8633	1277
7	7	127	101	12827	1437
....	....	....	....	....	....
10	10	997	839	836483	137101
10	10	839	967	811313	110349
10	10	787	967	761029	102957
....	....	....	....	....	....
512	512	See Fig-4	See Fig-4	See Fig-4	Large
....	....	....	....	....	....
1024	1024	See Fig-5	See fig-5	See Fig-5	Very Large

3 Comparative Result Analysis

After introducing Shor algorithm factorization attack is one of the major drawbacks over many classical cryptosystem like RSA. In RSA only one pair of keys (e, d) are used for encryption and decryption where as in our literature we have introduced set of encryption-decryption keys {e_1, e_2, e_3, e_4, e_5,..} and {d_1, d_2, d_3, d_4, d_5,..} respectively. Random selection of encryption key from the key set {e_1, e_2, e_3, e_4, e_5...} make the cipher non-deterministic and the corresponding decryption key is selected from the LUT (look up table) for decryption in every distinct run. Randomised key selection and non-deterministic cipher generation in every successive run makes our information secure. Although in 2019, RSA-240 (795 bits) [24, 25] and in 2020, RSA-250 (829 bits) [26] are factorized. Hence increasing bit patterns may vulnerable in next few years so as a result security may compromise. Now the question is how we can secure our data regardless of these threats? In our literature we have introduced characterise encryption using randomly selected encryption key one at a time from the encryption key set and maintaining a LUT which contains the corresponding decryption keys for each encryption keys. So hacking of only one decryption key is not sufficient for information leakage unlike RSA. To hack the complete information attacker must identify all the corresponding decryption keys for each of the encryption keys, and another issue is our generated large primes are very distinct hence encryption–decryption key set are also very distinct which generates different power trace on same plain text message in every run hence side channel attack may crucial.

4 Conclusion and Feature Work

Key generation and randomised key selection form the set of possible keys is our main concern. For same plaintext, different encryption key can use in different runs which makes the cipher distinct every time. Although same key can use for different characters in same plain text in a single run but two consecutive characters can never use same key for encryption because of randomised key selection. Also we have to notice our algorithm always produce distinct large prime numbers in every run which makes our key set non-deterministic. As a result randomised selection key form the non-deterministic key set makes our cipher non-deterministic. So an adversary can never track the power trace [11], timing variation from the cipher by using side channel attack [6, 7, 17, 18]. As the no of bits increases, the size of the key array also increases hence the possible candidates for key selection also increases exponentially this makes brute force attack very crucial. In future we ex tends our work by selecting only few candidate keys instead of considering all for encrypting plaint text of varying length to make the computation faster and more realistic.

Acknowledgments. I would like to convey my deep respect and sincere gratitude to my supervisor Dr. Souvik Bhattacharya (HOD), Dept. of CSE, University Institute of technology Affiliated under "The University of Burdwan". I also want say thank you to all of my teachers and colleague who have supported me a lot for doing my research work.

References:

1. Diffie W, Hellman M (1976) New directions in cryptography. IEEE Trans Inform Theor IT-22:644–654

2. Diffie W, Hellman M (1977) Exhaustive cryptanalysis of the NBS data encryption standard. Computer 10(6):74–84
3. Rivest RL, Shamir A, Adleman L (1978) A method for obtaining digital signatures and public-key cryptosystem. Commun ACM 21(2):120–126
4. Wiener MJ (1990) Cryptanalysis of short RSA secret exponents. IEEE Trans Inf Theor 36(3):553–558
5. Shor PT (1996) Polynomial-Time Algorithms for Prime Factorization and Discrete Logarithms on a Quantum Computer. arXiv: quant-ph/9508027v2, pp 1–28
6. Kocher P (1996) Timing attacks on implementations of Diffie-Hellman, RSA, DSS, and other systems. In: Koblitz Neal (ed) CRYPTO 1996, vol 1109. LNCS. Springer, Heidelberg, pp 104–113. https://doi.org/10.1007/3-540-68697-5_9
7. Kocher P, Jaffe J, Jun B (1999) Differential power analysis. In: Wiener M (ed) CRYPTO 1999, vol 1666. LNCS. Springer, Heidelberg, pp 388–397. https://doi.org/10.1007/3-540-48405-1_25
8. Schindler W (2000) A timing attack against RSA with the Chinese remainder theorem. In: Koç ÇK, Paar C (eds) CHES 2000, vol 1965. LNCS. Springer, Heidelberg, pp 109–124. https://doi.org/10.1007/3-540-44499-8_8
9. Mermin ND (2006) breaking RSA encryption with a quantum computer: Shor's factoring algorithm. In: Physics pp 481–681
10. Ghosh S, Alam M, Gupta IS, Chowdhury DR (2007) A robust GF (p) parallel arithmetic unit for public key cryptography. In: 10th Euromicro conference on digital system design architectures, methods and tools (DSD 2007), Lubeck, Germany, pp 109–115
11. Burman S, Mukhopadhyay D, Veezhinathan K (2007) LFSR based stream ciphers are vulnerable to power attacks. In: Srinathan K, Rangan CP, Yung M (eds) INDOCRYPT 2007, vol 4859. LNCS. Springer, Heidelberg, pp 384–392. https://doi.org/10.1007/978-3-540-77026-8_30
12. Rebeiro C, Mukhopadhyay D, Takahashi J, Fukunaga T (2009) Cache timing attacks on Clefia. In: Roy B, Sendrier N (eds) INDOCRYPT 2009, vol 5922. LNCS. Springer, Heidelberg, pp 104–118. https://doi.org/10.1007/978-3-642-10628-6_7
13. Sarkar S, Maitra S (2009) Partial key exposure attack on CRT-RSA. In: Abdalla M, Pointcheval D, Fouque P-A, Vergnaud D (eds) Applied Cryptography and Network Security: 7th International Conference, ACNS 2009, Paris-Rocquencourt, France, June 2-5, 2009. Proceedings. Springer Berlin Heidelberg, Berlin, Heidelberg, pp 473–484. https://doi.org/10.1007/978-3-642-01957-9_29
14. Sarkar S, Maitra S (2010) Cryptanalysis of RSA with more than one decryption exponent. Inf Process Lett 110(8–9):336–340
15. Sarkar S, Gupta SS, Maitra S (2010) Partial key exposure attack on RSA – improvements for limited lattice dimensions. In: Gong G, Gupta KC (eds) Progress in Cryptology - INDOCRYPT 2010. INDOCRYPT 2010. LNCS, vol 6498. Springer, Heidelberg. https://doi.org/10.1007/978-3-642-17401-8_2
16. Maitra S, Sarkar S, Sen Gupta S (2010) Publishing upper half of RSA decryption exponent. In: Echizen I, Kunihiro N, Sasaki R (eds) IWSEC 2010, vol 6434. LNCS. Springer, Heidelberg, pp 25–39. https://doi.org/10.1007/978-3-642-16825-3_3
17. Ghosh S, Mukhopadhyay D, Chowdhury DR (2011) Fault attack, counter measures on pairing based cryptography. Int J Network Secur 12(1):21–28
18. Banik S, Maitra S, Sarkar S (2012) A differential fault attack on the grain family of stream ciphers. In: Prouff E, Schaumont P (eds) CHES 2012, vol 7428. LNCS. Springer, Heidelberg, pp 122–139. https://doi.org/10.1007/978-3-642-33027-8_8
19. Rebeiro C, Mukhopadhyay D, Bhattacharya S (2015) An introduction to timing attacks. Timing Channels in Cryptography. Springer, Cham, pp 1–11. https://doi.org/10.1007/978-3-319-12370-7_1

20. Chakraborty A, Alam M, Dey V, Chattopadhyay A, Mukhopadhyay D (2018) Adversarial Attacks and Defences: A Survey. arXiv preprint arXiv: 1810.0006
21. Gidney C, Eker M (2019) How to factor 2048 bit RSA integers in 8 h using 20 million noisy qubits. arXiv: 1905.09749, pp 1–26
22. Bhatia V, Ramkumar KR (2020) An efficient quantum computing technique for cracking RSA using Shor's algorithm. In: 2020 IEEE 5th international conference on computing communication and automation (ICCCA), pp 89–94
23. Audio file digitization and encryption using ASCII conversion. In: International conference OPTRONIX 2016, vol 194, pp 489–495. SN–978-981-10-3907-2 (2017)
24. https://lists.gforge.inria.fr/pipermail/cado-nfs-discuss/2019December/001139.html
25. Boudot F et al (2020) Comparing the difficulty of factorization and discrete logarithm: a 240-digit experiment, 10 June 2020. https://eprint.iacr.org/2020/697
26. https://listserv.nodak.edu/cgi-bin/wa.exe?A2=NMBRTHRY;dc42ccd1.2002

Sustainable Smart Village Online Groundwater Level Monitoring System to Find the Recharging Capacity of Wells

Sapna Jain(✉) and M Afshar Alam

Jamia Hamdard, Delhi, India
{drsapnajain,aalam}@jamiahamdard.ac.in

Abstract. Groundwater is one of the world's generally disseminated, inexhaustible, and most significant assets. It is essential to see how much water is entering groundwater inventory, as this impacts how much water can be secured, taken from groundwater supplies for human use. Understanding the groundwater table is essential, and it helps the farmer to identify the suitable cropping pattern. For example, if the recharge is more than discharge, we can suggest long-duration and water-consuming crops like banana, sugarcane, turmeric. If the recharge is less and short duration and less water destroying crops like maize, vegetables, pulses farming can be cultivated. This paper discusses the overall hourly monitoring of the groundwater table using a sensor technology pressure sensor based on piezoelectric with a cable used with sensors in the open well. The adjustment in pressing factor from a sensor is straightforwardly corresponding to the well's tallness of water. The paper discusses that the online groundwater level monitoring system uses a versatile GSM portable APP, which causes the rancher to go for substitute editing design. The proposed online groundwater measurement system can be implemented and used for rural villages in India.

Keywords: Smart village · Groundwater · Sustainability · Recharge · Agriculture

1 Introduction

The evolution of the 2030 schedule for [1] multimillennial sustainable development goals described by the United Nations in 2015 emphasizes the existing improvement model's sustainability. Consequently, it needs to be applied in our Indian Villages to achieve Sustainable improvement goals (SDG). The SDG encompasses [2, 3] making cities and Human Settlements Inclusive, Safe, Resilient, and Sustainable. It recognizes that not the handiest sustainable control and improvement of the urban surroundings are essential for everyday existence, but also human settlements in rural regions and villages. Our county is the best consumer of groundwater within the international. India uses 25% of global groundwater. In India, natural groundwater recharges several methods are used.

The soil water balance method is a conventional method used to estimate groundwater recharge by estimating the inflow and water outflow. This method does not involve

R. Misra et al. (Eds.): ICIoTCT 2021, LNNS 340, pp. 47–56, 2022.
https://doi.org/10.1007/978-3-030-94507-7_5

physical representation from the field, and accurate calibration is required. The zero-flux plane method considers soil water storage below the zero-flux plane, and it is found to be an expensive method as it is challenging to locate a position of [4] zero flux plane. Chaturvedi's [5] empirical formulae, Amritsar and Krishna Rao, are preferred to find India's natural water recharge estimation. Though many practical methods exist, there is no system to see the current groundwater table level through online courses. There are no single comprehensive estimation techniques that give a reliable result. In India, Groundwater wells have increased drastically from 1960 to 2006. It indicates the recharging is less than the discharge level. As per reports, nearly [6] 50% of our country faces drought this year, especially Arunachal Pradesh, Tamil Nadu Gujarat though there is a normal monsoon in 2018 [7–10]. Recently, in Odisha, the farmers found themselves in poverty due to crop failure due to the worse climatic condition and decline in the groundwater table. So many farmers are committed suicide because of crop failure and poverty. The green climate fund, world bank, and Odisha state government helped install recharge shafts, but there are still no significant efforts in India to understand the groundwater recharge. In the Coimbatore district of Tamil Nadu, the groundwater table of Sulur and Annur blocks is found to be between 30 and 40 m below ground level, and farmers are finding it difficult in proper land use and cropping pattern. It is necessary to identify water recharging areas and construct recharge structures, says K. Mylswami of Siruthuli, an NGO involved in water conservation. Thondamuthur block falls under the "Dark" category, where groundwater utilization is more than 85%.

2 Literature Review

Smart agriculture services can help efficient and productive farming competencies among farmers [14–16]. Previously, farmers were growing the simplest Kharif plants. After recharging their bore wells, the farmers now cultivate rabi plants (wheat, gram) properly. The main issues faced by way of the agricultural regions are poverty, illiteracy, and unemployment. With the development of the era in the rural areas, agriculture improvement can study and generate employment opportunities [17]. The utilization and improvement of a period in the rural areas have more suitable the agricultural masses residing requirements and led to the development. The device shall contribute to achieving the sustainable development goal 2. The rural improvement of U.N. goals focuses on agriculture. For the development of agriculture, there was practical usage of innovative techniques, schemes, and methods, various styles of irrigation strategies, seeds, fertilizers, pesticides, and pesticides had been added in the farming procedures [11]. There has been an era in other areas and animal husbandry, rural cottage industries, fitness, strength, water control, rural housing, roads and communique, and rural schooling. The era has enabled the agricultural loads to set up conversation hyperlinks, carry out exchange and enterprise transactions, and generate attention, information, and statistics amongst the farming people.

When rural people attend educational institutions for a training session on operating technical machinery like computers, they'll feel apprehensive, but they're always keen and enthusiastic to find out [12, 13]. Thondamuthur may be a block located in the Coimbatore district in Tamil Nadu. Positioned within the populated area of Tamil Nadu, it's one of the 14 blocks of the Coimbatore district. As per the administration register, the block number of Thondamuthur is 365. The coalition has ten villages, and there's a complete of 18346 families during this block. The ground Water degrees from the 39-wide underground wells employed by TWAD during the post-Monsoon and pre-Monsoon time within the last five years. The typical floor water level in underneath floor level for pre-and post-monsoon as given in Fig. 1.

Jan 2013	May 2013	Jan 2014	May 2014	Jan 2015	May 2015	Jan 2016	May 2016	Jan 2017	May 2017	Jan 2018	May 2018	Jan 2019	May 2019	5 Years Pre Monsoon Average	5Years Post Monsoon Average
14.3	16.7	15.1	23.0	16.11	16.0	13.79	16.7	20.36	29.7	19.8	22.3	13.6	17.6	**16.1**	**20.3**

Fig. 1. Groundwater level

The groundwater level was 16.63. The groundwater level is in meters as per the TWAD board. This groundwater table is made by the system, and this information can be given to ranchers versatile through. In India, the groundwater table is estimated physically during the long periods of January, Pre-rainstorm, March/April/May, August, and Post-storm (November). It is recorded for a long haul to make an information base. This groundwater table put away in a disconnected information lumberjack requires field specialists to work with the product, and the instruments are cumbersome and expensive, surpassing multiple and half lakhs. In this paper, Thondamuthur Block in Coimbatore District identifies how the proposed Online Groundwater Level Monitoring System to find wells recharging capacity can improve the villagers' production and living conditions. Thondamuthur block is one of the overexploited areas. It has ten panchayat villages in which the primary crop cultivation are Groundnuts, Jowar (Rabi), Black gram (Rabi), Green gram (Rabi), Onion (Kharif), turmeric, banana. The installation of a water level measurement system in three to four open wells of three villages. The groundwater table measurement can be transmitted using cloud connectivity. Data recording regularly to estimate groundwater recharging capacity. This water management system helps locate areas with the rise or drop in groundwater table and estimate and forecast groundwater recharge. A smart village is a network in rural regions that leverages digital connectivity, solutions, and assets to improve and transform closer to achieving the desired results. The online Monitoring system uses a database and provides visualization of effects on the web portal. The three open wells selected from three to four villages in and around the Thondamuthur block of Coimbatore district shall help farmers improve the production of the crops grown in the area. The improvement in output shall increase income, which shall allow them to live a self-sustainable life.

3 Proposed System Design

The proposed method implementation will occur in the Thondamuthur block of the Coimbatore district of Tamil Nadu, India. Thondamauthur is an over-exploited area of groundwater. It has a total geographical area of 845.58 ha. It has ten panchayats villages like Devarayapuram, Ikaria,boluvampatti, Jagirnaickenpalyalam, madampatti, madvarayapuram, narasipuram, P.C. palayam, theethipalayam, thennamanallur as shown in Fig. 2. These villages are located west of District headquarters in Coimbatore. The Thondamuthur Town Panchayat has a population of 11,492, according to the census report 2019. Their primary crops are Groundnuts, Jowar (Rabi), Black gram (Rabi), Green gram (Rabi), Onion (Kharif), turmeric. Banana. Due to the decline in the groundwater table, water-consuming crops cultivated, and farmers find it difficult to estimate proper land use and cropping pattern. If the farmers know the groundwater table, they can predict the appropriate crop patterns. This information lumberjack framework helps the technocrats, specialists, and geohydrologists working in the Tamil nadu water management system know the reviving limit and over abuse of groundwater in specific areas. The plan shall monitor the groundwater level by measuring the water table in open wells. The microcontroller program shall get the water level fluctuations in the well using a pressure sensor.

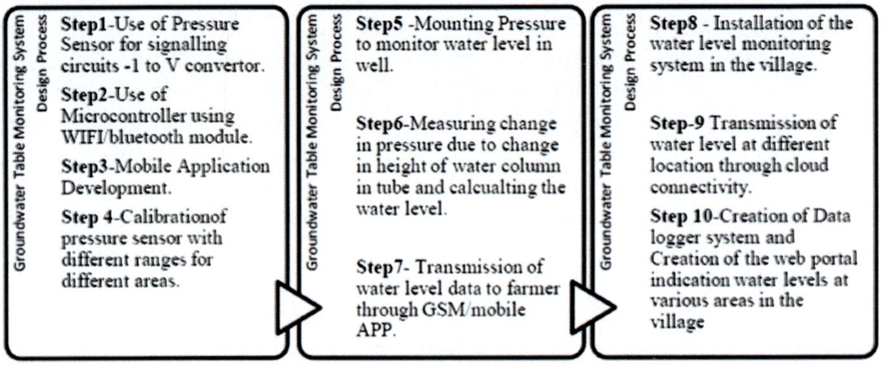

Fig. 2. Design steps of the proposed system

The procedure of proposed work comprises of three sections-

1) The reconnaissance survey has to be conducted to study the selected well-essential characteristics in the Thondamuthur block. An interview with farmers is to be done to get necessary information such as groundwater availability, rainfall, and land use and cropping pattern, etc.,
2) Design and Development of Groundwater table monitoring system.

Figure 2 shows the groundwater table monitoring system process for farmers. A pressing factor sensor alongside a link is embedded in the well till its tip. The adjustment in the critical factor is straightforwardly corresponding to the stature of water in the well. The adjustment in the tallness of the water segment inside the pipe changes the vital element. This critical factor correlates to the tallness of water present in the well by utilizing the equation

$$P = \rho gh \tag{1}$$

P is the crucial factor because of the water section in kg/m. s2.
ρ is the thickness of water = 1000 kg/m^3;
g is the quickening because of gravity = 9.81 m/s^2;
h = stature of water section in the line in m.

The groundwater table information is sent online through Bluetooth wi-fi network/GSM to the rancher for eco benevolent use. This aids in the legitimate recommendation about the appropriate trimming example to the ranchers relying on groundwater accessibility. The model is implemented in various tanks to recharge the groundwater table results using the standard ICT standard hardware.

3) Design and Development of Water Management System

The drawn-out information assists with assessing the groundwater energizing limit here. Incessant estimations should be possible hourly or regular routine. The proposed assessment is monetarily savvy, and periodic evaluation improves the exactness. Figure 3 shows the complete prototype model, which indicates the groundwater table for the farmers. It consists of a pressure sensor, current to voltage converter, microcontroller, ZigBee module, and LCD.

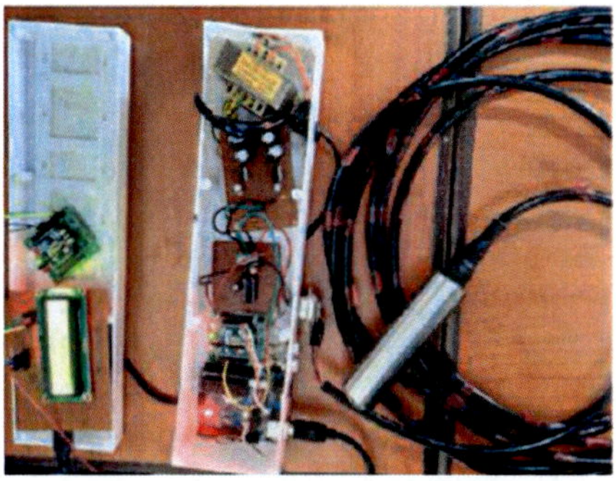

Fig. 3. Proposed system prototype

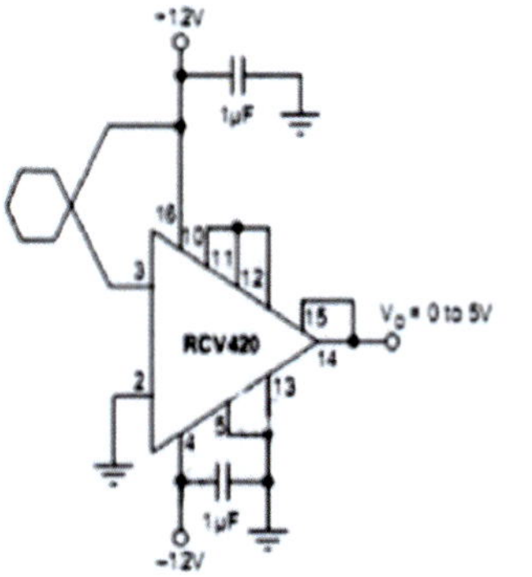

Fig. 4. Current to voltage convertor

The pressure sensor used is of submersible type. The output is 4–20 mA. This 4–20 mA current is converted into corresponding voltage levels of 0–5 V using RCV420 current to voltage convertor as shown in Fig. 4.

4 Results and Discussion

The Tamil Nadu nation is utilizing hydrogeological developments. Right around seventy-three% of the state is involved by using intense rocks. The semi merged lot arrangements are exceptionally confined inside the eastern segment, the waterfront plot. Inside the extreme stone area, groundwater is specifically developed through burrowed wells and burrowed cum bore wells tapping the endured area, the yield of open wells degrees from one to 3 lps. In contrast, in burrowed wells tapping soft rocks along the edge of sedimentary arrangements, the product is just about as much as five lps. Dynamic groundwater things have been surveyed at a couple of levels in the ebb and flow valuable asset's assessment yearly replenishable groundwater helpful support of the state has been anticipated as 22.94 bcm, and net yearly floor water accessibility is 20. sixty-five bcm [20]. When a year floor water draft is sixteen. 56 bcm and level of groundwater development is eighty% leaving limited degree for additional create ment of the powerful groundwater assets. Out of 386 assessment devices (blocks), 139 had been named over-taken advantage, 33 as imperative, 67 as semi-critical, 136 as quiet, and 11 as saline. Enormous groundwater improvement is seen inside the fundamental a piece of the U.S., and its miles presented out inside the classification map showing consideration of overexploited and basic square in a direct example broaden ing along the new way in the critical a piece. Groundwater furthermore should be considered as a piece of an included water resources control method that coordinates land and water sources. It oversees water sum and first-class linkages, manage surface water and groundwater assets conjunctively, supports tests, and reestablishes traditional designs. This included method offers new horrendous circumstances for groundwater control, consisting of the need for higher data of the results on groundwater recharge sum, fantastic of different floor water frameworks, and many interesting difficulties. The spring planning has added a solid realities base by incorporating the records acquired from geographical, hydrogeological, hydrological, geophysical, and geochemical considers. With the accessible use of the spring maps, there might be an entire evaluation of the groundwater accessible inside the province of Tamil Nadu and Puducherry as portrayed in Fig. 5. It might be done at appropriate scales by enveloping assorted regulatory units like squares, taluks, and companies of gram panchayats overlying on a spring. records of the spring and availability of the groundwater at the panchayat level might be fundamental for supported local area development on managing groundwater [18–20].

The proposed recharge borewells' effectiveness can be observed with data results accrued from wells in the vicinity's impact. The fluctuations were analyzed in advance than and after artificially recharging the aquifer. Water level fluctuations and water stability strategies had been used to quantify the person and blended effectiveness of various synthetic recharge systems in recharging the groundwater aquifer. The water stages elevated barely around 1.5 m at some point in July because of the southwest monsoon and are reduced in September. Due to rainfall in the northeast monsoon and stored water in test dams and recharging thru recharge borewells, the water stage turned into once more improved in all of the remark wells during December (5.7 mbgl). The recharge during the northeast monsoon for a duration up to December is about eleven. Four percent water stage fluctuation is observed with about 0.5 m during the southwest monsoon for a length up to september is ready sixteen. Eight in step with cent with typical water

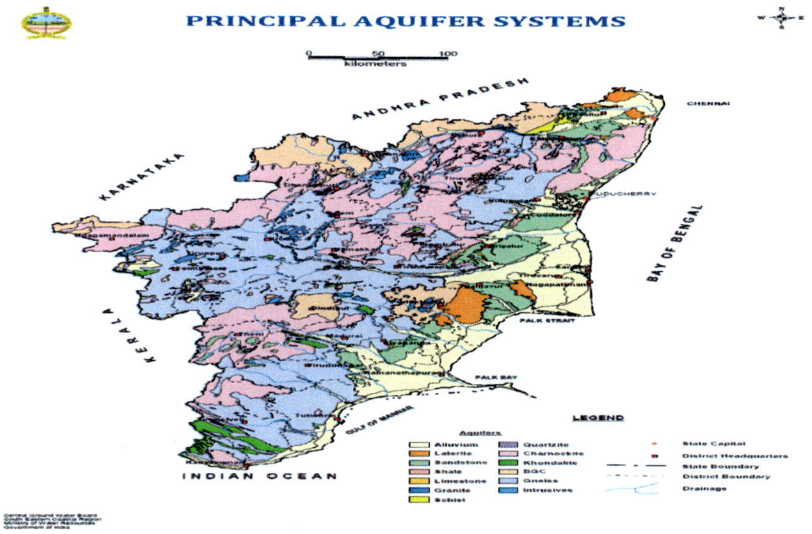

Fig. 5. [20] Water level fluctuation

level fluctuation of about 0.61 m inside the implementation region. Natural recharge, the intensity of groundwater growth is 1.5 m whereas the areas with artificial recharge structures increase the groundwater table's increase is four.7 m. The microcontroller reads the sensor information every time the clock pin is enabled.

Further, the sensor facts are transmitted using Xbee module. The evaluation end result of the groundwater table with the overall system is given in Table 1. The work shows that the software shall be helpful to farmers. The district-wise analysis using the ICT tools water recharge method in well is shown in Fig. 6.

Table 1. Proposed system impact

Measured ground water	Water table with standard equipment
10	10
20	20
30	30
40	41
50	51
60	61
70	71
80	81
90	91
100	101

The utilizable groundwater recharge is 22,423 MCM. The present utilization level is expressed as a net spring water draft of 13.558 MCM, which is set 60 in step with the available recharge. At the same time as 8875 MCM (40 consistent with cent) is that the stability to be had for use. over the past five years, the percentage of safe blocks has declined from 35.6 in keeping with cent to 25.2 according to cent at the same time as the semi-crucial blocks have long gone up by using the same percentage. Over-exploitation has already taken place in extra than a 3rd of the blocks (35.eight in step with cent) at the same time as eight blocks (2%) have grown to become saline. The water level statistics reveal that the depth of the wells tiers from a median of zero. Ninety-three meters in Pudukkottai district to 43.43 m in Erode.

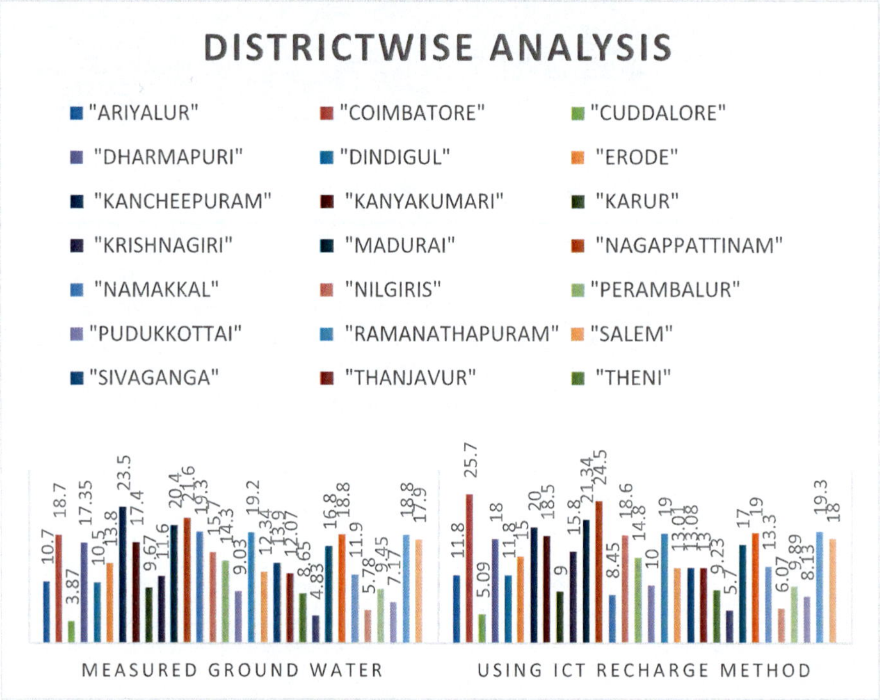

Fig. 6. Comparison districtwise analysis [21]

5 Conclusion

The proposed system shall contribute to the direction of the sustainable development goals and make Tamil Nadu self-sufficient. The proposed system implementation with the cooperation of experts and villagers. The water level assistance can help the farmers to improve the new possibilities of increasing agricultural cultivation. The problem of water scarcity is resolved efficiently throughout the year. Implementing a proposed system can help the platform of provincial zones innovation with a better approach to building their towns and making them up to a shining city. Smart Villages will make vocation simpler for the locals and a better future by using the latest technologies such as cloud computing and sensors technology for water management and farming.

Acknowledgement. The authors acknowledge the DST organization of the government of India for providing the Fund for Improvement of S&T Infrastructure (FIST) for the research lab in the Department of Science and Technology, Jamia Hamdard, to conduct this research work.

References

1. M Valeri (2019) Corporate social responsibility and reporting in sports organizations. CSR, sustainability, ethics & governance CSEG. Springer, Cham. https://doi.org/10.1007/978-3-319-97649-5
2. Verma S, Petersen AC (eds) (2018) Developmental science and sustainable development goals for children and youth, vol 74. SIRS. Springer, Cham. https://doi.org/10.1007/978-3-319-96592-5
3. Iyer-Raniga U (2018) Resetting the compass: principles for responsible urban built environment education (PRUE). In: Leal Filho W, Rogers J, Iyer-Raniga U (eds) Sustainable development research in the Asia-Pacific region. WSS. Springer, Cham, pp 31–77. https://doi.org/10.1007/978-3-319-73293-0_3
4. Jain SK (2012) India's water balance and evapotranspiration. Curr Sci 102(7):964–967
5. Kumar CP (1977) Estimation of natural ground water recharge. ISH J Hydraul Eng 3(1):61–74
6. Tripathi SS, Isaac RK (2016) Rainfall pattern and groundwater fluctuation in Ramganga Riverbasin at Bareilly District, Uttar Pradesh, India. Int J Adv Eng Manage Sci (IJAEMS) 2(6):239477
7. SyamRoy B (2017) India's journey towards sustainable population. Springer, Cham, pp 3–7. https://doi.org/10.1007/978-3-319-47494-6_1
8. Pirmana V, Alisjahbana AS, Hoekstra R, Tukker A (2019) Implementation barriers for a system of environmental-economic accounting in developing countries and its implications for monitoring sustainable development goals. Sustainability 11(22):6417
9. UN (2015). https://www.unep.org/explore-topics/sustainable-development-goals/why-do-sustainabledevelopment-goals-matter/goal-11
10. Akca H, Sayili M, Esengun K (2007) Challenge of rural people to reduce digital divide in the globalized world: theory and practice. Govern Inf Q 24(2):404413
11. Antrop M (2005) Why landscapes of the past are important for the future. Landsc Urban Plan 70(12):2134
12. JH Kim, RB Jackson (2011) A global analysis of groundwater recharge for vegetation, climate, and soils. Soil Sci Soc Am 11:vzj2011-0021RA
13. Oliveira PTS et al (2017) Groundwater recharge decrease with increased vegetation density in the Brazilian cerrado. Ecohydrology 10(1):e1759
14. Pérez-del Hoyo R, Mora H (2019) Toward a new sustainable development model for smart villages. In: Smart villages in the E.U. and beyond, pp 49–62
15. Visvizi A, Lytras MD (2018) It's not a fad: smart cities and smart villages research in European and global contexts. Sustainability 10(8):2727
16. Ward N, Brown DL (2009) Placing the rural in regional development. Reg Stud 43(10):12371244

17. Zavratnik V, Kos A, Duh ES (2018) Smart villages: comprehensive review of initiatives and practices. Sustainability 10(7):2559
18. Ground Water Year Book - India 2019-20, Central Ground Water Board
19. Ministry of Jal Shakti: Department of Water Resources, River Development and Ganga Rejuvenation, Government of India
20. Groundwater, the nature precious gift to mankind, lets manage properly. http://cgwb.gov.in/AQM/TamilNadu.pdf
21. https://www.winmeen.com/tamilnadu-police-constable-exam-cutoff-marks-result/?cv=1

Stacked Generalization Based Ensemble Model for Classification of Coronary Artery Disease

Pratibha Verma[1](✉), Vineet Kumar Awasthi[2], A. K. Shrivas[3], and Sanat Kumar Sahu[4]

[1] Department of Computer Science, Dr. C.V. Raman University Bialspur (C.G.), Bilaspur, India
[2] Department of Information Technology and Computer Science, Dr. C.V. Raman University Bialspur (C.G.), Bilaspur, India
[3] Department of Computer Science and Information Technology, Guru Ghasidas Vishwavidyalaya, Bilaspur, India
[4] Department of Computer Science, Government Kaktiya P.G. College Jagdalpur (C.G.), Jagdalpur, India

Abstract. Data mining based classification techniques plays an important role in medical data analysis that gives a better way to predict or diagnose any disease at an early stage. The development of a robust model is very important to achieve better classification accuracy. The proposed work constructed a robust ensemble model using a combination of Radial Basis Function Network (RBFN) and Random Forest (RF) with the stacking ensemble method. This research work has used two Coronary Artery Disease (CAD) datasets namely Z-Ali-Zadeh Sani (ZAZS) and Extension Z-Ali-Zadeh Sani (E-ZAZS) for analysis and check the robustness of models. It presents the robust ensemble model for predicting both datasets and showed improved accuracy over other traditional methods.

Keywords: Random Forest (RF) · Radial Basis Function Network (RBFN) · Ensemble model · Coronary Artery Disease (CAD)

1 Introduction

Coronary Artery Disease (CAD) develops by the formation of plaques inside the walls of coronary arteries, resulting in the narrowing of lumens of coronary arteries [1]. CAD is a serious problem for human life in the world. Nowadays much medical technology is available in healthcare which helps to improve the diagnosis of different diseases. The technological improvements in healthcare industries have facilitated accurate treatment in medical science but the treatment of CAD remains difficult for the common people. Nowadays, the healthcare organization produces huge amounts of complex and unformatted data concerning treatments of patients, diagnosis of disease, records of patients and resources of the hospital, etc. These large amounts of information can be processed and analyze using knowledge extraction tools which help for decision making as well as cost-saving [2]. The main objective of this study is to develop a robust model using data mining techniques that can handle the problem of CAD. Data mining is one of the important machine learning and knowledge discovery techniques that are used to find

R. Misra et al. (Eds.): ICIoTCT 2021, LNNS 340, pp. 57–65, 2022.
https://doi.org/10.1007/978-3-030-94507-7_6

the relevant patterns from large databases. In the healthcare industry, data mining plays a very important role in the identification and classification, and prediction of various diseases [3, 4]. Data mining provides many tools and technologies which are very useful to reach specific goals. In this research work, we have used RF, RBFN, and their ensemble model for the classification of CAD. In this research work, the two data sets as ZAZS and E-ZAZS have been used for analysis and check and compare the performance of models where our proposed ensemble model (RF + RBFN) gives better performance compared to the individual's classifiers. The proposed ensemble model quickly identifies and screens the CAD problems so that it will helpful for the early detection of CAD. This efficient model will be helpful for an enhanced patient caring system with restricted resources.

There are many researchers who have been working in the field of computer-based CAD diagnosis systems. This also includes heart diseases occurring in the human being. Javeed et al. [5] have suggested ANN and DNN with feature elimination technique FWAFE. The proposed hybrid techniques FWAFE-ANN were given the best 91.11% accuracy while FWAFE-DNN gives 93.33% accuracy. Baccouche et al. [6] have proposed various ANN techniques using ensemble learning techniques. The CNN with BiLSTM gives the best 91% accuracy with various types of heart disease. Latha et al. [7] have suggested bagging and boosting ensemble techniques with different base classifiers like Naive Bayes, Bayes Net, C4.5, MLP, and PART and also developed three ensemble models with a combination of these base classifiers using the stacking ensemble technique. Swain et al. [8] have proposed a Dense Neural Network for the classification of Cleveland Heart disease and achieved an accuracy of 94.91%, whereas 83.67% of accuracy was achieved by Miao et al. [9] by using the enhanced Deep Neural Network (DNN) for the classification of heart disease diagnosis. Trindade et al. [10] have used robust integrated bioinformatics tools with proteomic data and also studied similarities between CAD and Aortic valve stenosis. Babič et al. [11] have used different classifiers for the classification of healthcare datasets. They have also compared the performance of classifiers with different healthcare datasets and suggested the best classifier with the specific dataset. Bektas, Ibrikci, and Ozcan [12] have used different classification techniques like Logistic regression Neural Network, RNA with the cardiovascular dataset. They have also used feature selection techniques like Relief-F and Independent t-test analysis. They have suggested Neural Network gives the best accuracy with the Relief-F feature selection technique. El-bialy et al. [2] worked on the integration approach of machine learning techniques and analysis performs on CAD disease. The pre-processing operation is performing using this model and achieved satisfactory accuracy. Alizadehsani et al. [13] have suggested Naive Bayes, C4.5, and the K - Nearest Neighbor (KNN) classification technique for the classification of CAD.

2 Material and Methodology

The architecture of the proposed model is divided into different sections. The capability of the systems is estimated in terms of increasing the performance of classification accuracy. Figure 1 shows that the proposed architecture as explained.

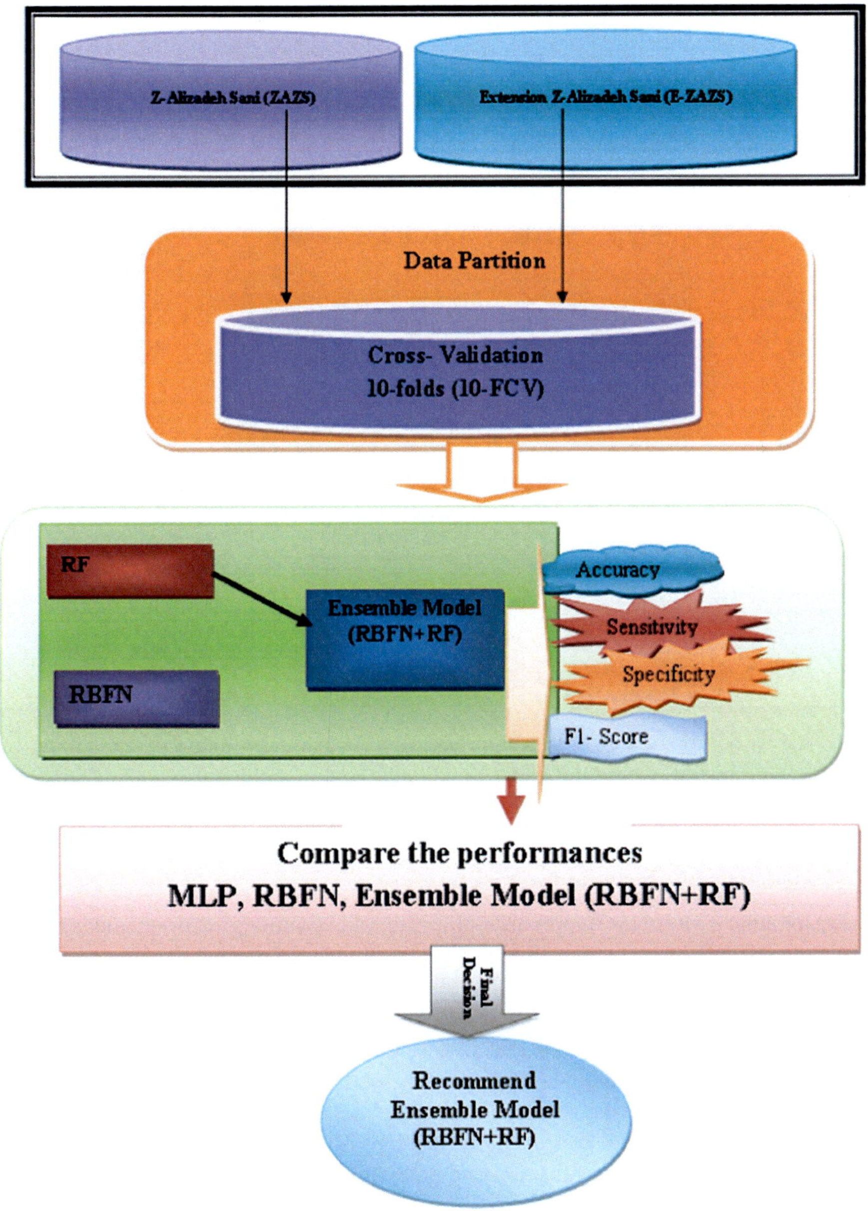

Fig. 1. Proposed architecture

2.1 Data Sets

Data set is very important for every experimental-based research work. This research work has used two different data set z-Ali Zadeh Sani (ZAZS) and extension of z-Alizadeh Sani E-ZAZS collected from the UCI repository. The description of both datasets is given Table1:

Table 1. Dataset description

Parameters	ZAZS dataset	E-ZAZS dataset
Number of features	56	59
Number of samples	303	303
Number of CAD patients	216	216
Number of normal patients	87	87
Missing value	No	No
Nature of class	Binary {Cad, Normal}	Binary {Cad, Normal}

2.2 Data Partition

K-fold cross validation is a technique for partitioning data into training and testing. This research work has used 10-fold cross validation for the partitioning of the dataset into training and testing where the dataset is divided into 10-fold and each fold is used as testing dataset then taken average accuracy of testing stage.

2.3 Classification Techniques

Classification is one of the important applications of data mining technique. This research has used RF, RBFN and its ensemble technique for classification of CAD.

- **Random Forest (RF):** RF is decision tree based on data mining and supervised machine learning technique. RF is an ensemble classifier that is combination of different decision tree technique to enhance the performance of model [15, 16]. The RF is capable to increase the accuracy if increase the number of trees in the forest.
- **Radial Basis Function Network (RBFN):** RBFN is a type of artificial neural network that is used for both classification and prediction. It consist only three layers: input layer, hidden layer and output layer. In the architecture of neural network, hidden layers nodes offers a collection of "functions" that represent an arbitrary "basis" for the input patterns after they are improved into the hidden space; these features are called RBF [17, 18].

– **Ensemble Method:** Data mining ensemble is the procedure of creating several models and combining them to produce the desired output. The main motive of an ensemble model that gives better performance compare than individual models. The focus this research work is to combining of different trained classifiers, hence increase the accuracy and efficiency of ensemble model. A stack Generalization is technique used to combine the two or more trained classifiers to make ensemble classifier. The output of ensemble classifier is input of next level meta classifier to learn the mapping between output of ensemble classifier and actual corrected classes.

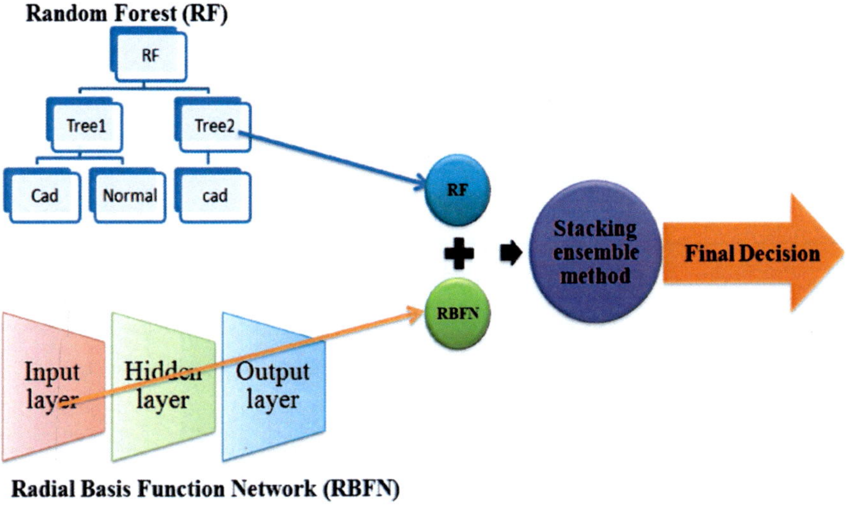

Fig. 2. A proposed ensemble model

Figure 2 shows an ensemble model, where two trained classification techniques is used to develop an ensemble model like RF decision tree and RBFN. The stacking generalization method used for developing an ensemble model.

Pseudo Code of Proposed Ensemble Model

```
Input: Input CAD dataset
Output: Acc=Accuracy
Function: Models (Classifiers)
        Dataset =CAD # Select the dataset (ZAZS or E-ZAZS)
        Folds = Cut (sequence (1, N row (CAD dataset)),breaks=10),
        While I<=10
        {
        do
        testIndexes = which(folds=I, arrayindex=TRUE)
        testData = CAD[testIndexes ]
        trainData = CAD[-testIndexes]
        }
        Choose Classifiers
        {
        Random Forest(RF), Radial Basis Function Network(RBFN)
        OR
        Ensemble_Stack()
        {
                Select Meta Classifiers RBFN
                 Models = [RF, RBFN]
                I, algorithm in enumerate (Models)
                {
                Classifiers [,I] = Models.fit(train, target).predict(test)
                }  }
        }
         Acc=Find Accuracy (CAD dataset, method).
         Return (Acc)
End
```

3 Classification Performance

The performance parameters like accuracy, Sensitivity, Specificity, F1-Score have been used to analyze and compare the robustness of classifiers. The various performance measures are calculated with the help of a confusion matrix [3]. Table 2 shows that confusion matrix that contains true positive (TP), true negative (TN), false positive (FP), false negative (FN).

Table 2. Confusion matrix.

Actual Vs Predicted	Positive	Negative
Positive	True Positive (TP)	False Negative (FN)
Negative	False Positive (FP)	True Negative (TN)

$$Accuracy = \frac{TP + TN}{TP + FP + FN + TN} \quad (1)$$

$$Sensitivity = \frac{TP}{TP + FN} \quad (2)$$

$$Specificity = \frac{TN}{TN + FP} \quad (3)$$

$$F1 - Score = \frac{2TP}{2TP + FP + FN} \quad (4)$$

4 Result and Discussion

The experiment is done in R data mining tools with a window 7 environment. The main focus of this research work is to develop a robust classifier that can classify the CAD. We have used ZAZS and E-ZAZS data sets for analysis and to check the robustness of models. The classifiers have been trained and tested using the 10-fold cross validation data partition technique. Table 3 shows the overall accuracy of RF, RBFN, and their ensemble where the proposed ensemble model achieved better accuracy than the individual models in the case of both ZAZS and E-ZAZS datasets.

Table 3. Accuracy of classifiers.

Name of algorithms	ZAZS dataset	E-ZAZS dataset
RBFN	79.21%	88.12%
RF	83.17%	94.39%
Proposed ensemble model	84.82%	96.04%

Table 4 shows that the F-score of individuals and proposed ensemble model where proposed ensemble model gives better accuracy as compared to individual classifiers in case of both datasets. The ensemble model has given the highest as 89.82% of F1-Score compared to individual models in the case of the ZAZS dataset. Similarly, the ensemble model has given the highest 97.23% of F1-Score compared to individual classifiers in the E-ZAZS dataset.

Table 5 shows that the proposed ensemble model has fluctuated with sensitivity performance measures as compared to individual classifiers like RBFN and RF. The sensitivity of RF is better as compared to other classifiers in the case of both datasets. The sensitivity of RF has given the highest sensitivity as 95.37% and 98.15% in the case of ZAZS and E- ZAZS datasets respectively.

Table 6 shows that the proposed ensemble model has fluctuated with specificity performance measures as compared to individual classifiers like RBFN and RF. The

Table 4. F1-score of classifiers.

Name of algorithms	ZAZS dataset	E-ZAZS dataset
RBFN	84.37%	91.39%
RF	88.98%	96.14%
RBFN + RF	89.82%	97.23%

Table 5. Sensitivity of classifiers.

Name of algorithms	ZAZS dataset	E-ZAZS dataset
RBFN	78.70%	88.42%
RF	95.37%	98.15%
RBFN + RF	93.98%	97.68%

Table 6. Specificity of classifiers.

Name of algorithms	ZAZS dataset	E-ZAZS dataset
RBFN	80.46%	87.35%
RF	52.87%	85.06%
RBFN + RF	62.07%	91.95%

specificity of RBFN is better as compared to other classifiers in the case of both datasets. The specificity of RBFN has given the highest specificity as 80.46% and 87.35% in the case of ZAZS and E- ZAZS datasets respectively.

The experimental results reveal that our proposed model gives better performance in the case of accuracy and F-score measures. Finally, the proposed ensemble model is recommended for the classification of CAD.

5 Conclusion

The classification of CAD is a very challenging and critical task for doctors and medical students. The researchers used different tools and techniques to enhance the performance of CAD identification and classification. This research work has proposed an ensemble model for classifying the CAD. This proposed model will be helpful for better deification and diagnosis of CAD in medical science. The stacking generalization method is used for developing the ensemble model where RBFN and RF classifiers are used to develop the ensemble model for classifying CAD in the case of both ZAZS and E-ZAZS datasets and also compared the performance. The proposed ensemble model gives 84.82% and 96.04% of accuracy in the case of both ZAZS and E-ZAZS datasets respectively. Finally, our proposed model is capable to detect CAD at an early stage and helps for diagnosis.

References

1. Pal D, Chakraborty C (2011) Data mining approach for coronary artery disease screening. In: 2011 international conference on image information processing, pp 1–6
2. El-bialy R, Salamay MA, Karam OH, Khalifa ME (2015) Feature analysis of coronary artery heart disease data sets. Proc. Comput. Sci. 65:459–468
3. Han J, Kamber M, Pei J (2012) Data Mining: Concepts and Techniques. Elsevier, Amsterdam
4. Pujari A (2013) Data Mining Techniques. University Press, Cambridge
5. Javeed A et al (2020) Heart risk failure prediction using a novel feature selection method for feature refinement and neural network for classification. Mob Inf Syst 2020:1–11
6. Baccouche A, Garcia-zapirain B, Olea CC (2020) Ensemble deep learning models for heart disease classification : a case study from Mexico. Inf 11:1–28
7. Latha CBC, Jeeva SC (2019) Improving the accuracy of prediction of heart disease risk based on ensemble classification techniques. Inf. Med. Unlocked 16:100203
8. Swain D, Pani SK, Swain D (2019) An efficient system for the prediction of coronary artery disease using dense neural network with hyper parameter tuning. Int. J. Innov. Technol. Explor. Eng. 8:689–695
9. Miao KH, Miao JH (2018) Coronary heart disease diagnosis using deep neural networks. Int J Adv Comput Sci Appl 9:1–8
10. Trindade F et al (2018) How to use and integrate bioinformatics tools to compare proteomic data from distinct conditions? A tutorial using the pathological similarities between aortic valve stenosis and coronary artery disease as a case-study. J Proteomics. https://doi.org/10.1016/j.jprot.2017.03.015
11. Babič F, Olejár J, Vantová Z, Paralič J (2017) Predictive and descriptive analysis for heart disease diagnosis. In: Proceedings of the federated conference on computer science and information systems, pp 155–163 https://doi.org/10.15439/2017F219
12. Bektas J, Ibrikci T, Ozcan IT (2017) Classification of real imbalanced cardiovascular data using feature selection and sampling methods: a case study with neural networks and logistic regression. Int J Artif Intell Tools 26:1750019
13. Alizadehsani R, Habibi J, Hosseini MJ, Boghrati R (2012) Diagnosis of coronary artery disease using data mining techniques based on symptoms and ECG features. Eur J Sci Res 82:542–553
14. Z-Alizadeh Sani dataset (2016) https://archive.ics.uci.edu/ml/datasets/extention+of+Z-Alizadeh+sani+dataset
15. Breiman L (1998) No Title. 1–15
16. Pal, M. (1996) Random forests for land cover classification Mahesh Pal department of civil engineering National Institute of technology. In: Kurukshetra, pp 3510–3512
17. Haykin S (2008) Neural Netwok: A Comprehensive Foundation. Pearson Prentice Hall, Hoboken
18. Zadeh M, Mahmoudi J (2017) Discrimination between earthquakes and explosions using artificial neural networks. Am. J. Biometrics Biostat. 2:1–6

Check for updates

Smart Waste Management System in Smart City

T. Jaya Sankar, A. Vivek, and P. C. Jain(✉)

Electrical Engineering Department, School of Engineering, Shiv Nadar University, Greater Noida, India
{tj258,va341,premchand.jain}@snu.edu.in

Abstract. Waste management has become one of the most important issues in our country. From densely populated cities to smaller rural communities, waste management systems keep our homes and communities free from unwanted clutter. The key to efficient waste management is to ensure useful segregation and disposal. The main problem in our country is the delay in the collection and disposal of waste. This is mainly due to the human intervention involved in the process of waste management. Smart Waste Management System deals with everyday problems of waste disposal by reducing this human intervention. In this paper we have implemented an efficient waste management system for the segregation and timely disposal of waste.

Keywords: Waste segregation · Waste disposal · Location · Motion detection

1 Introduction

In many cities, the municipal garbage bins are overflowing and they are not cleaned at proper time which results in overflow of garbage, and which again results in pollution, spread of diseases, and creates unhygienic conditions for the people, and ugliness to that place. A smart city aims to improve quality of life for its citizen by harnessing technology to connect infrastructure, resources, and services, for safer and more sustainable, livable, workable, and competitive living. Smart cities should be able to monitor and automate operations and applications in utilities, buildings, and infrastructure in real time. To meet these challenges, smart cities require a range of communication networks. Waste management is primary issue in smart city due to both the cost of the service and problem of storage of garbage in landfills. The use of intelligent waste container, which detects the level of waste load and allows for an optimization of the waste collector trucks route, can reduce the cost of waste collection and improve the quality of recycling. It can reduce waste collection costs by 30% and lowers carbon emissions by 60% as per M/S Sensoneo Solutions in U.S. Sensoneo also reported that its solution can achieve 30–60% reduction in waste collection route kilometres and 97% accuracy regarding actual waste production. The key to efficient waste management is to ensure useful segregation and disposal. The increase in waste generation due to rapid urbanization has led to serious problems like dealing with the methods of waste disposal. Effective disposal of waste has always posed a problem due to improper functioning of the respective municipal

R. Misra et al. (Eds.): ICIoTCT 2021, LNNS 340, pp. 66–72, 2022.
https://doi.org/10.1007/978-3-030-94507-7_7

authority. Currently, various garbage trucks have been kept in place to collect garbage regularly. But there are problems in this solution. This include people from the respective municipal authorities not showing up regularly to work or there might be a shortage of trucks i.e. assigning only a few trucks for various locations. All above leads to the delay in the collection and disposal of waste. The aim and scope of this work is to segregate dry-waste from the wet -waste and keeping a track of to what extent the garbage bin filled with dry and wet category. Then this data is communicated to the respective municipal authority. They can use this data for timely collection and disposal. A useful system comprising of GSM and GPS module has been included in this work to keep an accurate track of the garbage bins in various locations by communicating a warning message which includes location of the garbage bins also.

2 Related Work

Reference [1] implemented waste management using ultrasonic sensor connected to Arduino UNO to monitor waste bin garbage level. The event manager send alert to collect garbage using SMS in mobile phone when waste bin is full. Ref. [2] implemented using ultrasonic sensor, Arduino UNO micro-controller, and Wi-Fi module. The depth of the waste bin measured and the Wi-Fi module transmit above data to Internet. An open IoT platform Thingspeak is used to monitor garbage system. Ref. [3] implemented using Arduino, ultrasonic sensor, RF module. When waste bin full, sends a SMS using GSM/GPRS to mobile phone of truck driver to make it empty. Ref. [4] uses weight sensor and IR sensor so that it can provide effective solution to ultrasonic sensor. It also uses Wi-Fi module to get Internet connectivity. Ref. [5] demonstrates how IoT integration with data access network, geographic information system (GIS) can contribute to improve cities cleanliness management system. The system is based on IoT sensing prototype, which measures waste level of garbage bins and send data over Internet to a server for storage and processing. Ref. [6] developed smart garbage bin which alarms and inform the authorised person when the garbage bin is about to full. Ref. [7] used sensors to collect data from garbage bins by sending them to gateway using LoRa (Long range) technology which is low power and provides long range. The data from various garbage bins are collected by the gateways and sent to cloud over Internet using MQTT (Message queue telemetry transport) protocol. Ref. [8] is based on image processing and GSM module. The camera takes continuous snapshots of garbage bins. A threshold is set to compare output of camera and load sensor by which the weight of garbage is measured. If threshold level exceeds, micro-controller sends a message with the help of GSM module to collect garbage by local central office. Accordingly, authority sends garbage truck to collect the garbage. When 5G Cellular-IoT becomes popular, ultrasonic sensors can be embedded with NB (Narrow band)-IoT that monitors waste levels, and NB-IoT with GPS location can help to send real time data to municipality to optimize collection routes, pick up frequencies, and truck loads.

3 Smart Waste Management

In order to implement smart waste management, various methods have been adopted by different authors mentioned in above section. This paper proposed a solution to enhance waste collection efficiently using the Arduino micro-controller and ultrasonic sensor systems. Ultrasonic sensors use ultrasonic signals at frequencies above human hearing. The 40–70 kHz range is most popular. Its main aim is object detection and distance measurement. The ultra sonic transceiver measures time taken by the ultrasound signal to travel out to the target and back. Then knowing the speed of sound (1120 ft/s) in air the distance between ultrasonic transceiver and garbage can be find out. The ultrasonic sensor range is 8 to 10 m. In the proposed system using ultrasonic sensors, the overflow of garbage can be avoided and managed efficiently by communicating to the authorized person with the help of SMS or Email. This system also provides the information when the bin gets completely filled with garbage. This information reduces drastically the number of times the arrival of vehicle which collects the garbage by communicating with the authorities. This paper proposed a solution using a GSM module by which a garbage collector truck is able to keep a track of where the garbage is more, and requires immediate clearance. Then it will follow the route where the truck is first required to clean the garbage, without human monitoring. Ultrasonic sensors are used to check the level of waste and if the level of waste reaches the marked level, a clearance alert will be sent to the truck operators through the GSM module connected to the Arduino. It also portrayed the use of GPS module which is connected to the Arduino to be able to send location updates to the truck operator. This proved to be an efficient way of utilizing the available resources and keeping track of the level of waste and deploying required methods at the same time.

4 Methodology

We have implemented Smart Waste Management System utilizing a network of sensors, GSM and GPS modules, and micro-controller. The system was divided into two parts: One comprising of all the required sensors to be able to collectively work together for the segregation of waste and obtain the useful data. The second part deals with methods to how one can communicate this obtained data to the respective municipal authority.

4.1 Waste Segregation Implementation

This part mainly focuses on how to segregate waste effectively. Figure 1 shows a smart waste dustbin that has two ultrasonic sensors, a moisture sensor, a stepper motor, a PIR sensor, a GSM module and a GPS module. Initially, the dustbin is divided into two equal halves by inserting a cardboard. Two ultrasonic sensors are mounted on top of the dustbin in both halves, one for the dry category and the other for the wet category.

A PIR sensor is attached on the top right so as to detect the motion of a person nearby. A stepper motor is also fixed at the top left of the dustbin and the input to the stepper motor is the signal obtained from the moisture sensor placed underneath of the cardboard plate near stepper motor. This cardboard plate is created with a hole in the centre to accommodate the moisture sensor. This plate is attached to the handle of the stepper motor. When the motion is detected by the PIR sensor, the moisture sensor becomes active. As the user drops the waste on the plate, the moisture sensor measures the moisture of waste dropped and sends out the signal to activate the stepper

Fig. 1. Smart waste dustbin attached with sensors, GPS, and GSM module

motor. Accordingly, the stepper motor would now tilt the plate left or right based on the moisture contained in the waste dropped. Figure 2 shows respective block schematic of smart waste management system using Arduino micro-controller board, two ultrasonic sensors, a PIR sensor, a moisture sensor, a stepper motor, a GSM module, and a GPS module.

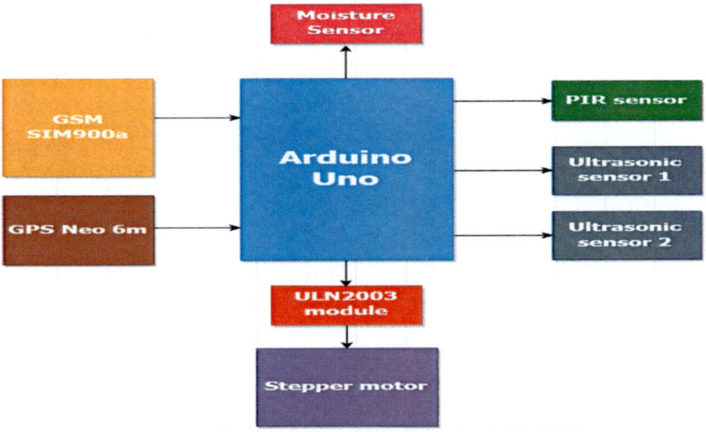

Fig. 2. Block schematic of smart waste management system

4.2 Transmission of Warning Message to the Respective Authority

The second step is to communicate the warning message to the respective authorities. When the dustbin gets filled up with dry or wet category or both, GSM module transmits a warning message to a respective municipal authority person's mobile phone stating the location of the dustbin using the GPS module and a warning message explaining that the dustbin is filled and requires immediate clearance.

5 Results

The ultrasonic sensors measure the level of waste, and accurately keep the track of the level of the waste in the dustbin as shown in Fig. 3. When the ultrasonic sensors keep monitoring the level, the PIR sensor detects whether there is any motion nearby. As soon as it detects motion, the moisture sensor is activated and sends out the signal to activate the stepper motor based on the waste dropped on the plate which is attached to the stepper motor. Based on the moisture values (Fig. 4), the waste is classified into either dry- or wet-waste and dropped to the corresponding section of the garbage bin.

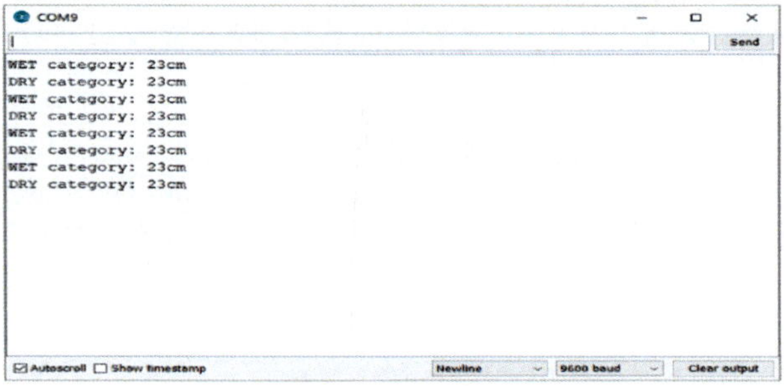

Fig. 3. Ultrasonic sensors continuously checking the level of waste

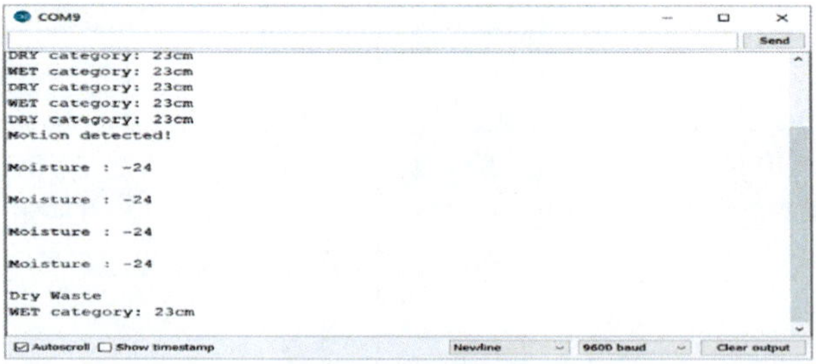

Fig. 4. Moisture sensor measures the moisture of the waste dropped

When the dustbin gets filled (ultrasonic sensors values drop below a threshold level), a warning message needs to be transmitted stating that the dustbin is full along with the location of the dustbin (Fig. 5).

As soon as the GPS gets the location of the dustbin, a message with a Google maps link is sent to the authorized person's mobile phone through the GSM module (Fig. 6).

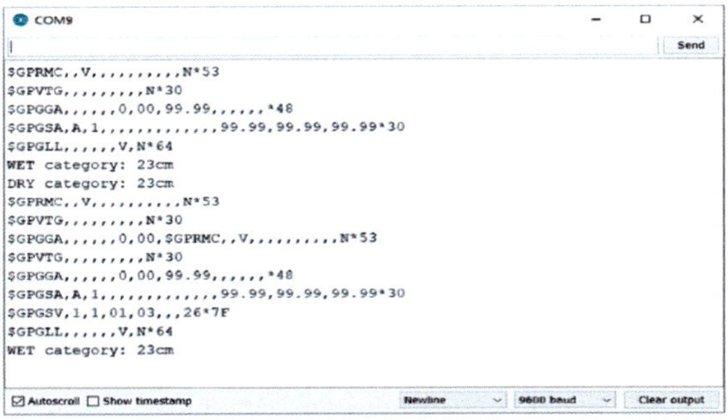

Fig. 5. GPS trying to acquire the location of dust bin

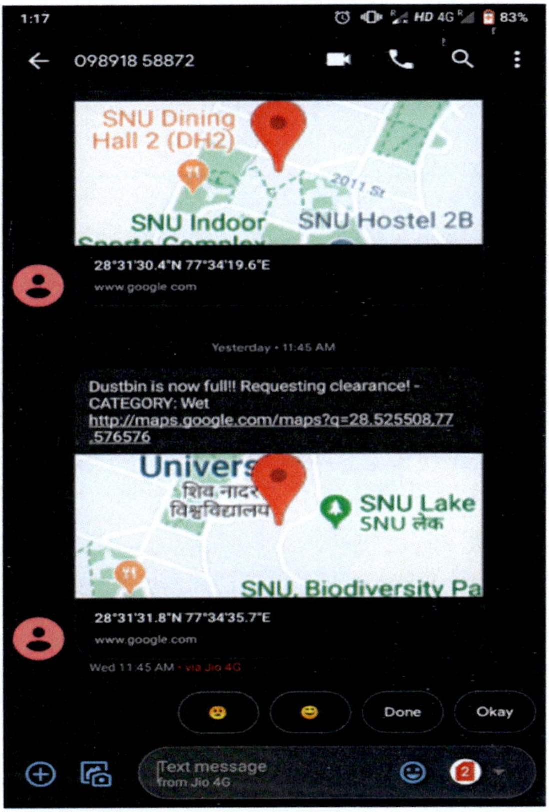

Fig. 6. Warning message on google map

6 Conclusions

Main objective of this paper is to achieve cleanliness in the cities and provide pollution free environment for better living. The aim behind implementing this system is to save time and effectively utilize the resources which are allocated for the waste management in the cities. Presently a few resources are available to process the waste management, and these resources are not utilized properly due to lack of communication. Our system aims to deploy a method that could help in better utilization of above resources. For example, if a single garbage collector truck is allotted to ten different areas, it has to move to each and every area on a daily basis. But with our system, the truck driver is notified when the dustbins get filled along with the location, thus allowing him to directly reach that specific area instead of just going through all the areas even though the dustbins aren't even filled. The next improvement can be made by upgrading the software. Instead of just sharing the location of the filled dustbin to the truck driver, one can devise an algorithm which shows the driver the shortest path to reach the dustbin. Internet of Things (IoT) and cloud combined with our implemented system, the waste management of the whole city can be managed from a single control point in the municipality.

References

1. Soh ZHC, Husa MAH, Abdullah SAC, Shafie M.A (2019) Smart waste collection monitoring and alert system via IoT. In: 2019 IEEE 9th symposium on computer applications and industrial electronics (ISCAIE), pp 50–54
2. Saha HN, Gon S, Nayak A, Moitra S (2018) IoT Based Garbage Monitoring and Clearance Alert System. In: 2018 IEEE 9th annual information technology electronics and mobile communication conference (IEMCON), pp 204–208
3. Hassan SA, Jameel NGM. Şekeroğlu B (2016) Smart Solid waste monitoring and collection system. Int J Adv Res ComputSci Softw Eng 6(10): 1–12
4. Navghane SS, Killedar MS, Rohokale VM (2016) IoT based smart garbage and waste collection bin. Int. J. Adv. Res. Electr. Commun. Eng. (IJARECE) 5(5):1576–1578
5. Gutierrez JM, Jensen M, Henius M, Riaz T (2015) Smart waste collection system based on location intelligence. Proc. Comput. Sci. 61:120–127
6. Thakker S, Narayanamoorthi R (2015) Smart and wireless waste management. In: 2015 international conference on innovations in information embedded and communication systems (ICIIECS) pp 1–4
7. Bharadwaj AS, Rego R, Chowdhury, A (2016) IoT based solid waste management system: a conceptual approach with an architectural solution as a smart city application. In: 2016 IEEE annual India conference (INDICON), pp 1–6
8. Prajakta G, Kalyani J, Snehal M (2015) Smart garbage collection system in residential area. Int. J. Res. Eng. Technol. (IJRET) 4:122–124

Carbon Rate Prediction Model Using Artificial Neural Networks (ANN)

Arunabh Bhattacharjee(✉) and Somnath Chattopadhyaya

Indian Institute of Technology, Dhanbad, India

Abstract. The work describes a data mining prediction model for control of carbon rate in a blast furnace iron making process. The proposed model not only enhances the control on fuel consumption, but also contributes to the thermal stability of the blast furnace. The carbon rate consists of coke rate plus coal rate and is measured per ton of hot metal. It is continuously monitored through measurement of the chemical composition of the gas exiting the furnace top known as "top gas". This paper presents an advanced data mining model that uses the technique of artificial neural networks (ANN) which can accurately predict the carbon rate with the key factors impacting the process. The accuracy of the model largely depends on the volume and the quality of data. Further, the paper reconfirms that artificial neural networks (ANN) is a better prediction technique than the conventional regression methods.

Keywords: Carbon rate · Data mining · Artificial Neural Networks · Machine learning · CRISP-DM · Blast furnace · Regression · Activation function

1 Introduction

A blast furnace iron making operation is highly complex and non-linear in nature. Besides this, due to the current competitiveness in the steel industry, it has become very essential to reduce the cost of production of hot metal. The fuel that is used to reduce iron ore accounts for approximately 40–60% of the total cost of hot metal. The most widespread form of fuel that is used for reducing iron ore into hot metal is the metallurgical coke. The importance of metallurgical coke in the blast furnace further rests in its ability to maintain bed permeability and structural integrity under the reducing conditions as well as the load of the burden.

The process efficiency of the blast furnace, generally considered to be the reductant (coke plus injectants containing carbon) rate per ton hot metal, and is continuously monitored through measurement of the chemical composition of the gas exiting the furnace top known as 'top gas'. The efficiency is expressed as the gas utilization formula, which indicates the percentage of CO gas utilized for reduction of ferrous burden and is calculated as:

$$\eta_{CO} = \frac{CO_2}{CO + CO_2}$$

R. Misra et al. (Eds.): ICIoTCT 2021, LNNS 340, pp. 73–87, 2022.
https://doi.org/10.1007/978-3-030-94507-7_8

It, henceforth, becomes necessary to develop systems that enhance control on fuel rate to reduce cost and, simultaneously, contribute to the thermal control and operational stability of the blast furnace.

To circumvent this, the present work introduces the concept of advanced data mining model, that uses the technique of artificial neural networks (ANN) to predict the carbon rate in the reduction process of iron making. This predictive model very efficiently utilizes the enormous data that is generated in an iron making process and, finally, affects the carbon rate or fuel consumption. The accuracy of the prediction, mainly, depends upon the input parameters, the volume of data used and their sensitivity towards the fuel/carbon rate of the process. To analyze this sensitivity of each input parameters towards the carbon rate, the machine learning technique of neural networks was also utilized to carry out a correlation analysis that, finally, gave the knowledge of the most significant and sensitive variable that affects the carbon rate in a blast furnace operation.

2 Related Work

Data mining is a powerful technique to extract useful, unknown, and actionable information from large volumes of data. Any data mining project follows a standard methodology called CRISP DM (Wirth and Hipp 2000). The CRISP DM is an internationally recognized methodology which follows a step by step method to use data mining to solve any business problem. This is a tool independent and industry neutral methodology.

Data mining can also focus to be domain specific. Domain specific data mining can be of real interest to business users (Longbing and Chengqi 2006). Of all the data mining prediction techniques artificial neural networks (ANN) is unarguably the most powerful technique (Lu et al. 1996). With the proposed approach, concise symbolic rules with high accuracy can be extracted from a neural network. The network is first trained to achieve the required accuracy rate (Lu et al. 1996). The effectiveness of the proposed approach is clearly demonstrated by the experimental results on a set of standard data mining test problems. First the carbon rate prediction is done using conventional regression techniques and then using ANN. It was found that prediction using ANN was closer to actual. Similar work was done on Comparison of linear regression and artificial neural network model of a diesel engine fueled with biodiesel-alcohol mixtures (Tosun et al. 2016). Application of artificial neural networks can be seen in other areas of Ironmaking like iron ore agglomeration (Dwarapudi et al. 2006) and coke making (Vasudharini et al. 2018). Data mining and artificial neural networks technique is not limited to ironmaking and steelmaking but has wide application in other areas like utput power of wind turbines (Tabatabaei 2017). It is also used for prediction of shape defects over length of cold rolled sheet (Raju et al. 2007). The paper on data driven detection of counter pressing in professional football matches presents how combination of positional and event data can result in a supervised machine learning model (Bauer and Anzer 2021).

3 Methodology

The research follows the CRISP-DM methodology. CRISP-DM, which stands for Cross-Industry Standard Process for Data Mining, is an industry-proven way to guide the data mining efforts. As a methodology, it includes descriptions of the typical phases of a project, the tasks involved with each phase, and an explanation of the relationships between these tasks. As a process model, CRISP-DM provides an overview of the data mining life cycle. The life cycle model consists of six phases with arrows indicating the most important and frequent dependencies between phases (Fig. 1). The sequence of the phases is not strict.

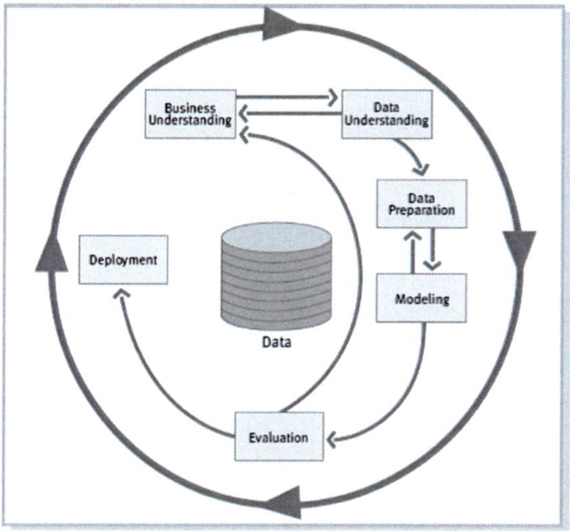

Fig. 1. Phases of CRISP methodology

In fact, most projects move back and forth between phases as necessary. The CRISP-DM model is flexible and can be customized easily. CRISP-DM allows to create a data mining model that fits your needs. Majority of time in data mining life cycle goes in data understanding and data preparation (Fig. 2).

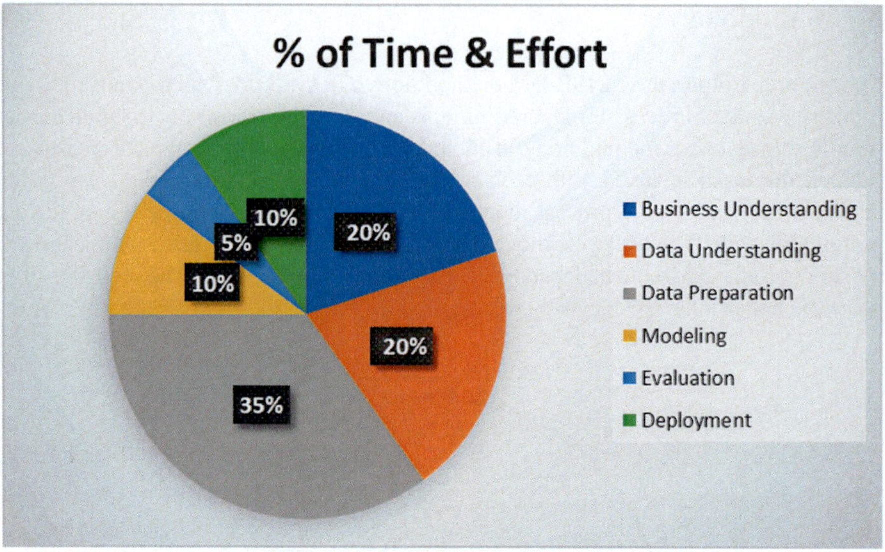

Fig. 2. Time & effort -DM life cycle

4 Data Preparation

The data preparation phase covers all activities needed to construct the final dataset [data that will be fed into the modeling tool(s)] from the initial raw data. Total volume of data used for this exercise is around 10000. In data preparation, all the parameters were converted to same frequency to form a square matrix. This is a very important step and all the parameters should be converted to the frequency of majority of parameters.

As a next step, the data subset is cleaned with removal of outliers and missing values. All erroneous values discarded to get a clean subset of data. Most data mining tools do not accept if the data set contains any error in the form of null values, special characters etc. Given below is the list of the parameters with description and unit of measure (UOM) used for analysis (Table 1).

Table 1. Data descriptions

Parameters	Unit of measurement	Description
CARBON_RATE	Kg/thm	Coke Rate + Coal rate
PRODUCTION	Tons	Hot metal production
AVAILABILITY	%	Ratio of available hours to total hours
COKE_RATE	Kg/thm	Coke consumed per Ton of Hot Metal Produced
COAL_RATE	Kg/thm	Coal consumed per Ton of Hot Metal Produced
FUEL_RATE	Kg/thm	Fuel consumed per Ton of Hot Metal Produced
SLAGRATE	Kg/thm	Slag discharged per Ton of Hot Metal Produced
ORE_RATE	Kg/thm	Ore consumed per Ton of Hot Metal Produced
SINTER_RATE	Kg/thm	Sinter consumed per Ton of Hot Metal Produced
PYROXINITE_RATE	Kg/thm	Pyroxenite consumed per Ton of Hot Metal Produced
O2RATE	Kg/thm	Oxygen consumed per Ton of Hot Metal Produced
HL_TOTAL_GJ_HR	GJ	Heat Loss (GJ)
SINTER%	Kg/thm	Sinter consumed as % of total Iron bearing material
PELLET%	Kg/thm	Pellet consumed as % of total Iron bearing material
PELLET B2	%	Pellet basicity
HMT	Degree Centigrade	Hot metal temperature
RAFT	Degree Centigrade	Race- away adiabatic flame temperature
HBT	Degree Centigrade	Hot Blast Temperature
ETA_CO	%	Gas utilization
HUMIDITY	%	Humidity in air
COKE ASH	%	Ash% in Coke
COKE VM	%	Volatile matter (VM) in Coke
COKE CSR	%	Coke strength after reaction (%)
MOIST LOAD (KG/THM)	Kg/thm	Total moisture load

Kg/thm: Kg per ton of hot metal.

5 Analysis

5.1 Carbon Rate Impact on Blast Furnace

This initial phase focused on understanding the project objectives and requirements from a business perspective, then converting this knowledge into a data mining problem definition and a preliminary plan designed to achieve the objectives. As a first step, blast furnace operation and all the key factors impacting carbon rate were measured on correlations matrix (Fig. 3). Coke rate, fuel rate, production, slag rate etc. were found to have high positive correlations with Carbon rate.

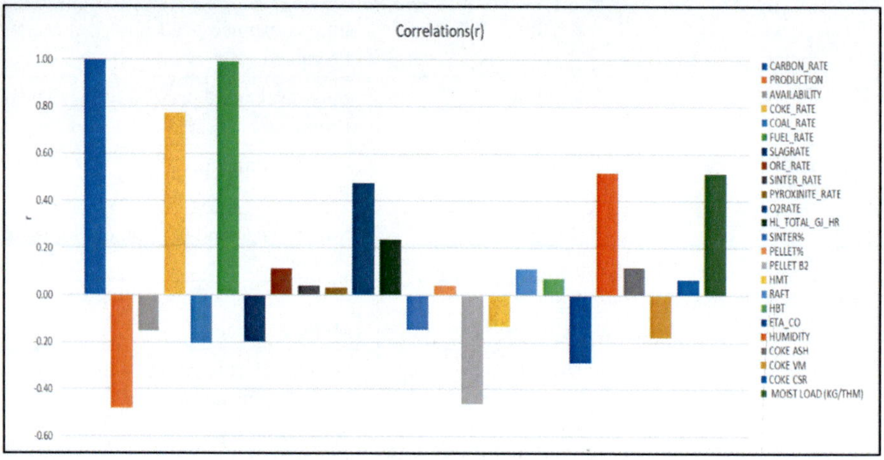

Fig. 3. Correlations with C-rate

Finally, the degree of impact was represented by a control impact matrix (Fig. 4).

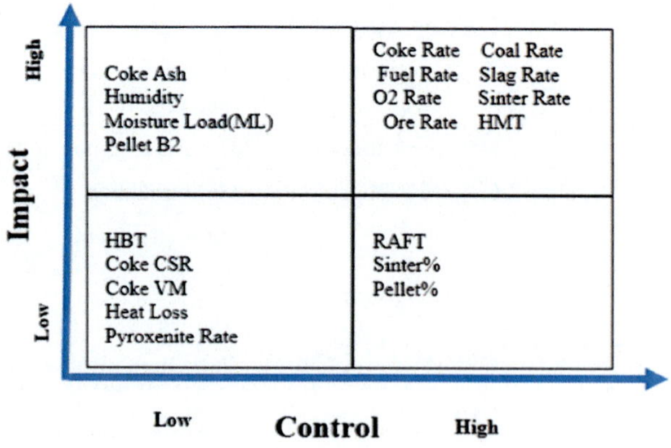

Fig. 4. Control impact matrix

5.2 Modeling

In this phase, various modeling techniques are selected and applied, and their parameters are calibrated to optimal values. Typically, there are several techniques for the same data mining problem type. Some techniques have specific requirements on the form of data. Therefore going back to the data preparation phase is often necessary. Two different approaches were followed to predict carbon rate

1. Using multiple regression
2. Using artificial neural networks (ANN)

Table 2. Regression summary

	b*	Std. err of b*	b	Std. err of b	t(811)	p-value
Intercept			1919.982	238.6724	8.04443	0.000000
Slag rate	0.00329	0.031949	0.003	0.0269	0.10293	0.918045
Ore rate	−2.09006	0.332350	−0.834	0.1325	−6.28872	0.000000
Sinter rate	−1.18508	0.461681	−0.480	0.1870	−2.56689	0.010427
Pellet rate	3.31403	0.717255	1.168	0.2528	4.62044	0.000004
Quartz rate	−0.01922	0.027991	−0.068	0.0986	−0.68682	0.492378
Pyroxinite rate	−0.04180	0.027803	−0.124	0.0827	−1.50330	0.133122
Limestone rate	0.03307	0.024441	0.234	0.1733	1.35317	0.176353
O2 rate	0.13087	0.031023	0.022	0.0052	4.21854	0.000027
RAFT	0.08789	0.026193	0.051	0.0152	3.35539	0.000827
Humidity	0.10838	0.031174	0.318	0.0915	3.47678	0.000533
Pellet B2	−0.06701	0.024555	−232.887	85.3412	−2.72889	0.006483
Pellet RI	−0.00898	0.023387	−0.112	0.2927	−0.38400	0.701070
Sinter RI	−0.01790	0.022962	−0.209	0.2683	−0.77961	0.435831
CSR	0.01882	0.023255	0.229	0.2829	0.80947	0.418464
Sinter%	−0.75476	0.144975	−5.924	1.1380	−5.20610	0.000000
Pellet%	−5.37752	1.033832	−32.162	6.1832	−5.20154	0.000000
Sinter Fines%	0.05957	0.026685	0.529	0.2371	2.23244	0.025838
Ore Fines%	0.00960	0.025638	0.077	0.2046	0.37451	0.708117
Pellet Fines%	0.09130	0.026597	0.631	0.1839	3.43265	0.000626
Heat Loss	0.12896	0.026742	0.096	0.0200	4.82247	0.000002
Overall ML	0.38842	0.035146	0.922	0.0834	11.05148	0.000000

The present research then compares how acurately both of these methods can predict Carbon Rate.

5.3 Prediction with Multiple Regression

Thus the entire problem can be reduced to y = f (x1, x2, x3.......xn). ThusC-Rate = f (process parameters, raw materials). The regression summary is given below (Table 2).

The scatterplot of predicted vs actual carbon rate is showing a linear behavior (Fig. 5).

5.4 Testing the Regression Model

Regression Summary for Dependent Variable: Carbon Rate R = .74199283 R^2 = .55055336 Adjusted R^2 = .53974193,F (21,873) = 50.923 $p < 0.0000$ Std.Error of estimate: 12.479. Therefore final equation can be derived as

*C−Rate = 0.003 * Slag Rate−0.834 * Ore Rate −0.480 * Sinter Rate + 1.168 * Pellet Rate−0.068 * Quartz Rate −0.124 * Pyroxinite Rate + 0.234 * Limestone Rate + 0.022 * O2 Rate + 0.051 * AFT + 0.318 * Humidity −2322.887 * Pellet B2 –0.112 * Pellet RI −0.209 * Sinter RI + 0.229 * CSR−5.924 * Sinter% −32.162 * Pellet% + 0.529 * SinterFines% + 0.077 * OreFines% + 0.631 * Pellet Fines% + 0.096 * Heat Loss + 0.922 * Overall ML + 1920*

The plot of C-Ratepredicted with C-RateActual when tested with new data points (Fig. 5).

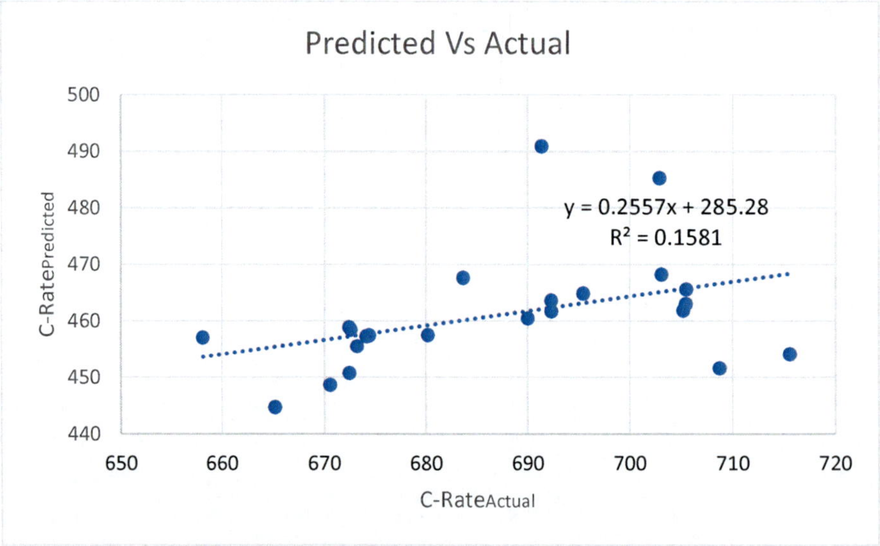

Fig. 5. Precited vs actual C-rate with linear regression

5.5 Prediction with Artificial Neural Networks

In this case all key parameters impacting C-Rate are taken as the input for the model. All machine learning models are self-learning and mature with more volume of data. Building a model using machine learning techniques follows 6 steps (Fig. 6).

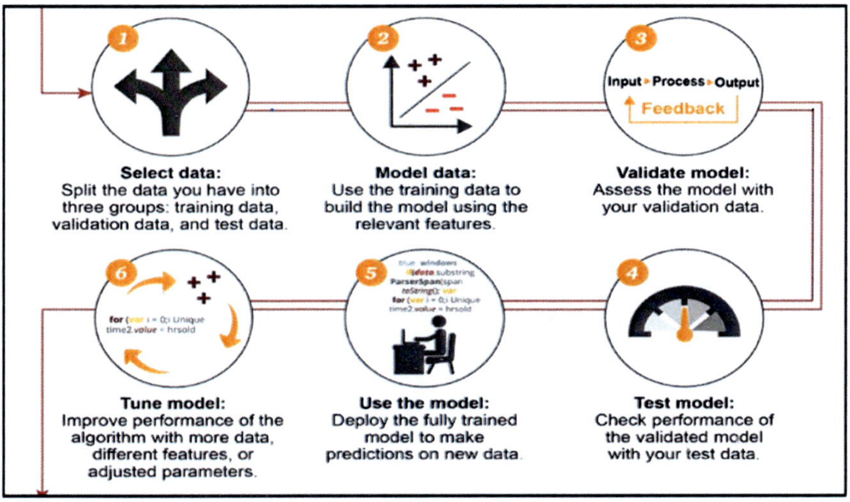

Fig. 6. How machine learning algorithm works

An artificial neural networks (ANN) works on the principle of self-training. The neural network algorithms are very calculation intensive. The sweeping success of neural networks over almost every other statistical technique can be attributed to power, versatility and ease of use. Neural networks are very sophisticated modeling and prediction making techniques capable of modeling extremely complex functions and data relationships. Successful applications in wide range of problem domains as diverse as finance, research, engineering, geology, stock market prediction etc. The activation function is a combination of transfer and activation function (Fig. 7).

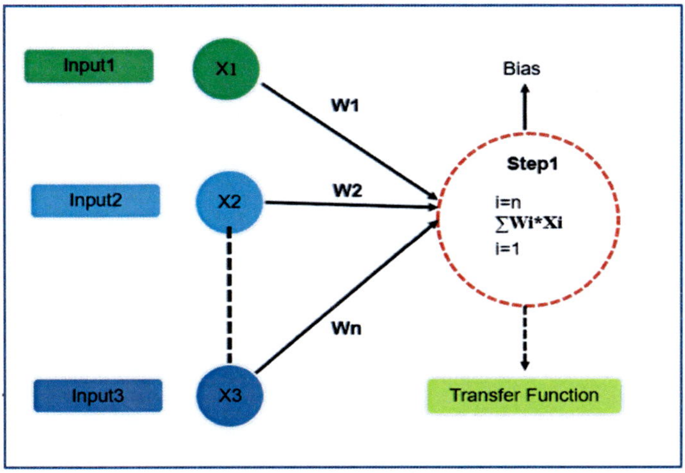

Fig. 7. A typical neural network function

An artificial neural network consists of several parameters and hyperparameters that drive the output of a neural network model. Some of these parameters are weights, biases, number of epochs, the learning rate, batch size, number of batches, etc.

Each node in the network has some weights assigned to it. A transfer function is used to calculate the weighted sum of inputs and a bias is added.

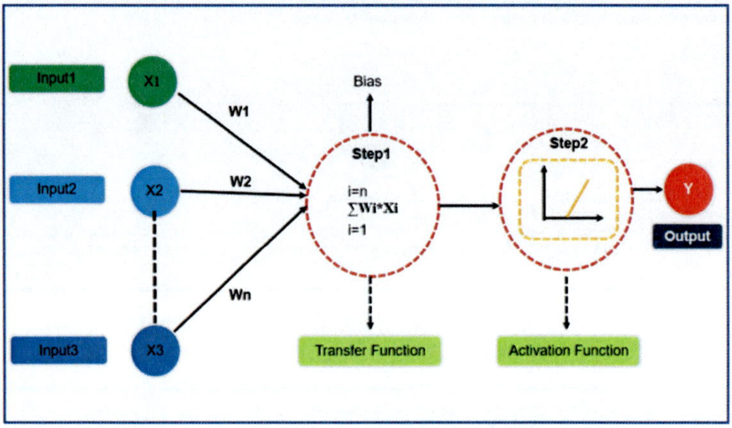

Fig. 8. Transfer & activation function of neural networks

The result of the transfer function is fed as input to activation functions (Fig. 8). Activation functions decide which nodes to fire. There are various activation functions that are used for specific purposes based on the type of output you are looking for. Some of the activation functions are Sigmoid, Step (Threshold), ReLU (Rectifier), Softmax, Hyperbolic Tangent function, etc. All these activation functions have specific usages. Some the functions are used only in the hidden layers, some in the output layers and there are a few which are used both in the hidden and output layers. Based on the fired nodes, the output layer produces the final predicted output. But, there might be errors in the predicted output or the predicted output might be varying too much as compared to the original output.

A Neural network model also has some pros & cons as described here (Table 3).

Table 3. Neural networks pros & cons

Pros	Cons
• Can learn more complicated class boundaries • Fast application • Can handle large number of features	• Slow training time • Hard to interpret • Hard to implement: trial and error for choosing number of nodes

The data set is divided into two subsets of data i.e. training set & test set. The training set is used to create the model and the test set is used to validate the model. Artificial neural networks (ANN) is subject of separate research. However, practicing engineers and data scientists worldwide use the power of ANN in commercially available packages or codes. These packages have inbuild artificial networks code so data scientists just need to use these packages with the input data to get the desired output.

The trend of the predicted C Rate with actual C Rate closely resemble each other (Fig. 9). Sample R code to predict carbon rate is given below (Fig. 10).

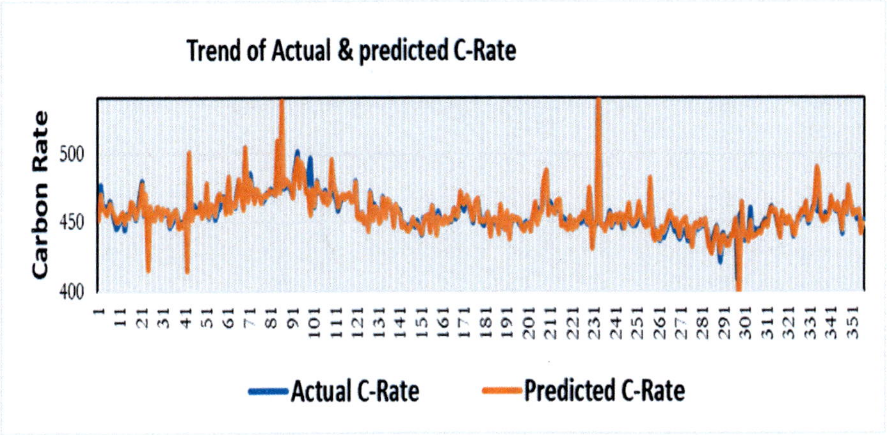

Fig. 9. Trend of predicted vs actual C rate

The neural network (NN) plot (Fig. 11) shows 3 nodes that are formed in the hidden layer.

6 Evaluation

Before proceeding to final deployment of the model, it is important to thoroughly evaluate it and review the steps executed to create it, to be certain the model properly achieves the business objectives. From considering both the options, ANN appears to have greater accuracy in predicting the Carbon Rate from the given data set. Though neural networks are a versatile technique that works well with any kind of noisy data it is not a guarantee that every time neural networks will have a higher accuracy. At the end of this phase, a decision on the use of the data mining results should be reached.

For a given data set the predicted vs actual is very close of the order of R^2 –0.8 (Fig. 12). For the model to remain relevant and updated, it should be trained with fresh data from time to time. Thus, the artificial neural network model (ANN) emerges to be a more powerful model to predict Carbon Rate.

7 Deployment

After the model is validated, the model is moved to production by implementing a scoring system where the model is applied to new data that doesn't have a dependent variable.

```
Neural Networks Prediction
# Random sampling
samplesize = 0.60 * nrow(df)
set.seed(80)
index = sample( seq_len ( nrow ( df ) ), size =
samplesize )
# Create training and test set
datatrain = df[ index, ]
datatest = df[ -index, ]
## Scale data for neural network
max = apply(df , 2 , max)
min = apply(df, 2 , min)
scaled = as.data.frame(scale(df, center = min, scale
= max - min))
## Fit neural network
# install library
install.packages("neuralnet ")
# load library
library(neuralnet)

# creating training and test set
trainNN = scaled[index , ]
testNN = scaled[-index , ]
# fit neural network
set.seed(2)
NN=neuralnet(CARBON_RATE~PRODUCTION+AVAILABILITY+COAL
_RATE+SLAGRATE+ORE_RATE+SINTER_RATE+PELLET_RATE+QUART
Z_RATE+PYROXINITE_RATE+LIMESTONE_RATE+OXYGEN_RATE+RAF
T+HUMIDITY+PELLET_B2+PELLET_RI+SINTER_RI+CSR+SINTER_P
ERC+PELLET_PERC+SINTER_FINES+ORE_FINES+PELLET_FINES+H
L+ML, data=trainNN, hidden = 3 , linear.output =
T,threshold=0.01)
# plot neural network
plot(NN)
## Prediction using neural network
predict_testNN = compute(NN, testNN[c(-4)])
predict_testNN = (predict_testNN$net.result *
(max(df$CARBON_RATE) - min(df$CARBON_RATE))) +
min(df$CARBON_RATE)
```

Fig. 10. Sample R code to predict carbon rate

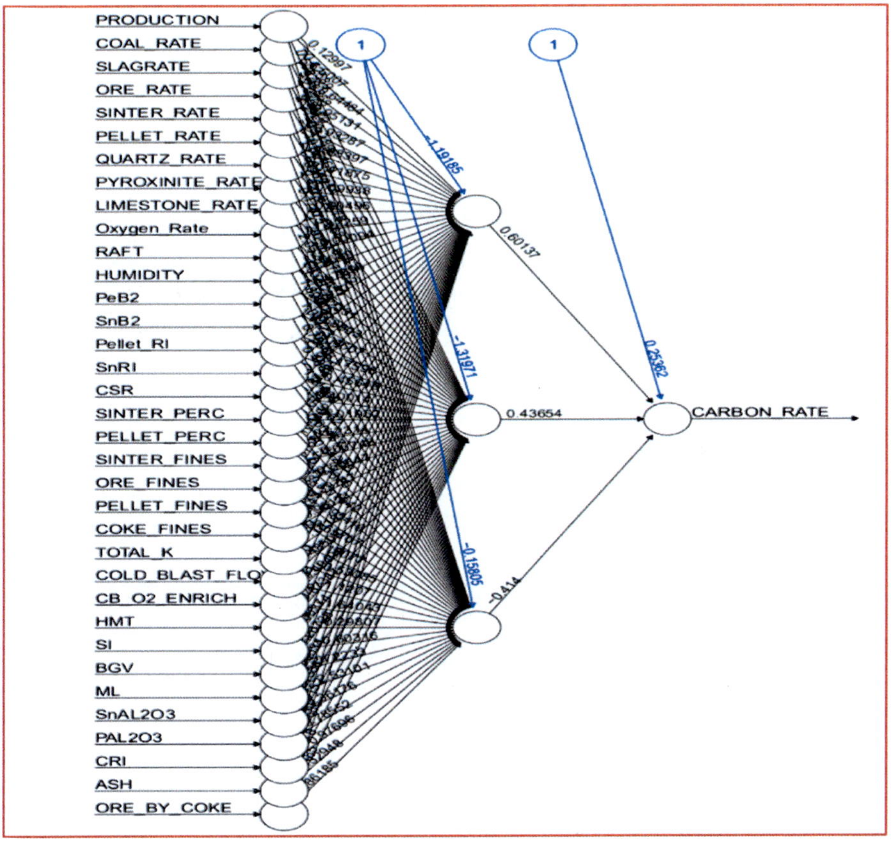

Fig. 11. NN plot in R

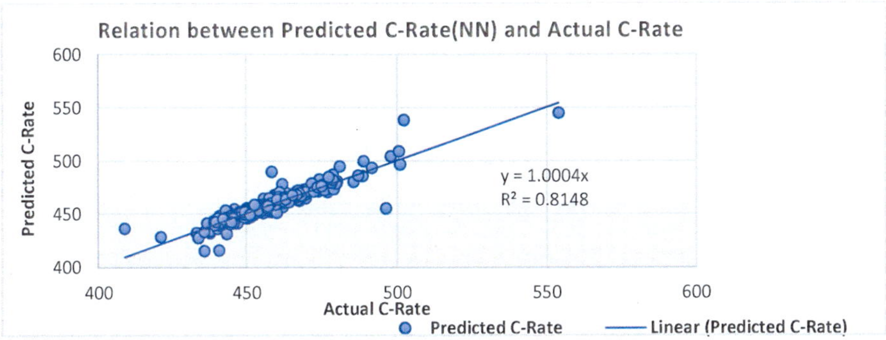

Fig. 12. Precited vs actual C-rate

This may require fine tuning the model and sometimes revisit the entire data mining process used to build the model. There must be a visualization platform so that front line officers can use it for running blast furnace operation. Multiple inputs from plant operators will make it a more robust model.

8 Conclusions

This paper presents a very systematic approach to predict key blast furnace parameters using powerful machine learning techniques like artificial neural networks (ANN). The neural network prediction of blast furnace carbon rate is very accurate as compared to the technique of multiple regression. It largely depends on the nature of the data. Even though the paper deals with Carbon rate prediction the technique can be used to predict other key parameters in blast furnace iron making as well as in other areas as diverse as sinter and coke making.

9 Future Scope

- The model accuracy and the concept need to be tested with data from other blast furnaces of varying capacities
- Further study can be done to understand the key parameters that increases the variability of the model.
- Can this kind of model run in real time at blast furnace control rooms so that the operators can accurately predict the Carbon rate and take corrective and preventive action.

References

Wirth R, Hipp J (2000) CRISP-DM: towards a standard process model for data mining. In: Proceedings of the 4th international conference on the practical applications of knowledge discovery and data mining. Citeseer

Cao L, Zhang C (2006) Domain-driven data mining: a practical methodology. Int J Data Warehousing Mining 2(4):49–65. https://doi.org/10.4018/jdwm.2006100103

Lu H, Setiono R, Liu, H (1996) Effective data mining using neural networks. IEEE Trans Knowl Data Eng 8(6):957– 961. https://doi.org/10.1109/69.553163

Raju DV, Iqbal A, Kumar Trivedi A, Mukhopadhyay A (2007) Prediction of shape defects over length of cold rolled sheet using artificial neural networks. Ironmaking Steelmaking 34(2):166–176. https://doi.org/10.1179/174328107X155268

Dwarapudi S, Gupta PK, Gupta SS (2006) Application of artificial neural network model to predict reduction degradation index of iron oxide pellets. Ironmaking Steelmaking 33(6):500–506. https://doi.org/10.1179/174328106X94807

Vasudharini SV, Dash PS, Singh RK (2018) Prediction of carbonisation time in byproduct recovery coke ovens through gooseneck temperature measurement. Ironmaking Steelmaking. https://doi.org/10.1080/03019233.2017.1398501

Tosun E, Aydin K, Bilgili M (2016) Comparison of linear regression and artificial neural network model of a diesel engine fueled with biodiesel-alcohol mixtures. Alex Eng J 55(4):3081–3089

Tabatabaei S (2017) ' A probabilistic neural network-based approach for predicting the output power of wind turbines. J Exp Theor Artif Intell 29(2):273–285. https://doi.org/10.1080/0952813X.2015.1132272

Bauer P, Anzer G (2021) Data-driven detection of counterpressing in professional football. Data Min Knowl Disc 35(5):2009–2049. https://doi.org/10.1007/s10618-021-00763-7

Check for updates

An Internet of Things Powered Model for Controlling Vehicle Induced Pollution in Cities

Mohamed Fazil Mohamed Firdhous[1,3](✉) and Janaka Wijesundara[2,3]

[1] Department of Information Technology, University of Moratuwa, Moratuwa, Sri Lanka
firdhous@uom.lk

[2] Department of Architecture, University of Moratuwa, Moratuwa, Sri Lanka
jwijesundara@uom.lk

[3] UoM Urban Lab - Center for Cities, University of Moratuwa, Moratuwa, Sri Lanka

Abstract. In recent years, the road traffic has increased in many cities creating many problems including increased pollution. Professionals and policy makers have time to time come up with schemes for controlling traffic congestion in cities. These schemes have met both successes and failures depending on many factors. The application of ICT has improved the efficiency and effectiveness of almost all the fields. In transportation, applications like Google maps help drivers to identify best routes to their destination based on the traffic conditions on different road segments. No ICT application including Google maps takes pollution as a factor when identifying the best route. This paper presents an IoT powered scheme for managing traffic at junctions based on the pollution levels at identified locations. The results obtained from tests conducted using simulations show that pollution at specific locations in cities can be controlled using this scheme.

Keywords: Data science · Internet of Things · ICT model · Congestion · Environment pollution · Mobile applications

1 Introduction

During the last few decades, the number of vehicles on roads has increased rapidly almost in every country [3,4]. This increase of vehicles has created many problems including increased number of accidents, traffic congestion, pollution of land, air and water, waste of natural resources such as fossil fuels, increased health related issues, and reduced productivity. One of the most serious issues that demand the immediate attention of the policy makers, planners, implementers and even the users is the effect of increased traffic on the environment. The pollution caused by vehicles has both short term as well as long term effects on people, animals as well as vegetation in the affected areas [9]. The effect is even more serious when the environment degradation happens in places where vulnerable groups of people like young children, older people and patients live. If the problem of traffic congestion

R. Misra et al. (Eds.): ICIoTCT 2021, LNNS 340, pp. 88–97, 2022.
https://doi.org/10.1007/978-3-030-94507-7_9

is not handled properly, it will not only increase the individual travel costs, but also result in the collapse of the entire municipal transportation system restricting urban sustainable development. Thus, solving traffic congestion problem has become a hot issue for almost every city.

With the development of advanced technologies many new methods for solving problems can be developed. For managing the traffic problem also, experts have developed many solutions including adaptive traffic signals, electronic toll collections and vehicle information communication service [2,7]. Applications like Google Maps, Waze and MapQuest have become popular among drivers for obtaining driving directions [11]. The main shortcoming of these applications is that they provide the directions based on the traffic conditions on road segments only. By following the suggestions given by these applications, it is possible to reach the destination faster probably reducing the amount of fuel burnt and emissions. Though, indirectly this helps reducing the environment pollution in an area they do not specifically target the reduction of environment pollution. Hence, these application stop short of reducing the environment pollution especially in vulnerable areas such as near schools, hospitals and other residential areas where the vulnerable groups of people live or congregate. Hence it is required to come up with methods, techniques and applications that can manage traffic more effectively considering a wide range of parameters. In this paper, the authors present an ICT model leveraging IoT, data science and other advanced techniques for managing traffic in cities with special emphasis on mitigating the impact of traffic on the environment.

This paper is organized into six sections as described here. Section 1 gives an overall introduction to the paper along with a brief discussion on the effects of traffic congestion. Sections 2 briefly describes the research methodology adopted in this work and Sect. 3 presents the critical review of the related work published in the literature. Section 4 presents the ICT model for environment aware traffic management proposed in this work. Section 5 presents the implementation details along with the experiments carried out for testing the system, the results obtained and a brief discussion on them. Finally, Sect. 6 concludes the paper with a discussion on how to extend this work in a future study.

2 Research Methodology

In order to carry out the proposed work in a scientific manner, an agile research and development methodology was devised and used. Figure 1 shows the main steps involved in the devised methodology.

The research process started with an extensive literature review on environment aware traffic management schemes proposed in the literature. During this process, it was found that not many environment aware traffic management schemes have been proposed solidifying the gap in this area requiring further in depth research. Once the research gap has been properly identified and solidified, the following research question has been formulated.

RQ: How to develop an ICT model incorporating the recent advances in the field for mitigating the negative environment effects caused by increased traffic in cities?

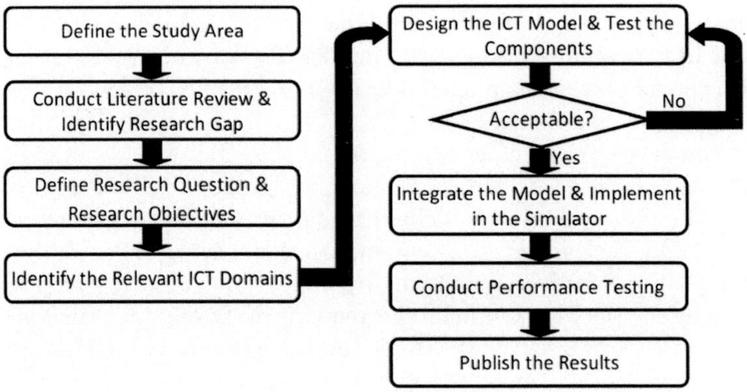

Fig. 1. Research methodology

Then, the specific objectives were defined:

O1: To develop an ICT model incorporating the recent advances in the field for mitigating the negative environment effects caused by increased traffic in cities.
O2: To implement the model in a simulated environment.
O3: To test and validate the model to ensure that it works satisfactorily.

Then, the recent developments in the field of ICT were studied in depth with special emphasis on their ability to collect information about the environment, process them effectively producing tangible results and then to disseminate the results to intended parties. Thus, Internet of Things (IoT), Cloud Computing (CC), Data Science (DS), Vehicular Communication with special emphasis on Vehicle to Infrastructure Communication (V2IC) and Mobile Application Development were studied in depth. Once, the study of the relevant ICT domains were completed, the ICT model was designed and each component of the model was analyzed in depth for functionality. Once the functionality testing was completed, the ICT model was created by integrating the components together. As at the current stage, it is not possible to test the system using field implementations or test-bed implementations, the implementation was carried out on SUMO (Simulation of Urban MObility). SUMO is an open source, highly portable, microscopic and continuous multi-modal traffic simulation package that is suitable for even designing and conducting large scale traffic simulation in complex urban settings [1]. Finally, the performance of the proposed model was tested using different cases.

3 Related Work

This section briefly surveys the work carried out by other researchers on the use of ICT in traffic management with special emphasis on environmental impact mitigation. Huang et al., have in [5] proposed an IoT based wireless sensor system for traffic volume and vehicle classification. They have implemented a prototype unit using wireless accelerometers for testing purposes. The prototype unit was installed on a shoulder next to a road without interfering with the ongoing traffic monitoring the pavement acceleration. The monitored pavement acceleration was used determine the number of vehicle axles, axle spacing and speed. The proposed unit is capable of determining the different groups of vehicles passing at a specific point based on the number of axles and axle spacing. The downside of this system is that it has only very limited application in managing the impact of the vehicular traffic on the environment.

Matharia and Dave have in [10] presented a smart traffic management system using IoT with the assistance of vehicle mounted RFID tags. The vehicle mounted passive RFID tags identify each vehicle uniquely to the roadside mounted RFID readers. When a vehicle passes a roadside RFID reader, the RFID tag mounted on the vehicle gets activated and transmits information including its ID to the reader. The reader, in turn transmits it to the central control along with a time stamp. This information will be used to compute the vehicle's speed and direction. When the over speeding vehicles were detected, the driver of the vehicle and the highway patrol are alerted. The proposed system is very simple and is capable of monitoring vehicle movements based on their speeds only. Hence, it does not have any application in managing traffic with respect to the maintaining a livable environment near highways or other roadways.

IoT based traffic management system for smart cities proposed by Sowmya et al. in [12] have the ability to dynamically control traffic flow by onsite traffic officers using smart phones or centrally using the Internet. The control of traffic flow is achieved through dynamically changing the signal timings at intersections. The algorithm used in this system for managing traffic employs a single parameter and does not take any other information such as type of vehicles or environmental parameters into consideration for making decisions.

The IoT based adaptive traffic management system proposed by Lalitha and Pounambal alters the signal times based on the number of vehicles traveling in a given direction [6]. The proposed mechanism employs an IR sensor to count the number of vehicles traveling in a given direction and use that information to alter the signal times with the objective of reducing congestion at junctions. This is a very primitive approach and will not reduce the cumulative waiting time of all the vehicles or number of vehicles going through a given point. Hence, this approach has very limited application in managing pollution at critical locations.

Malik et al., have in [8] described a methodology for efficiently managing traffic in highly populated and congested areas using IoT. The proposed framework employs decision algorithm and round-robin algorithm to find the optimum path through traffic. Infrared sensors are used for determining total traffic in a given region and the RFIDs attached to the vehicles enable system to detect high priority vehicles such as ambulances, fire trucks etc., for routing them fast. The main downside of this proposal is the requirement of an RFID tag attached to every vehicle. The authors also claim that the RFID will help to detect robbed vehicles. When a vehicle is robbed, the first thing the robbers would normally do is disabling or removing such equipment from the vehicle. Hence, the proposed methodology has very limited application in real world scenarios.

From the above discussion it can be seen that traffic management and pollution monitoring have not been researched together by any researcher so far. Hence, there is a huge need to look at them together arriving at practical solutions. The authors of this paper expects that this paper would be a starting point towards the development of integrated traffic and pollution management systems using a single ICT platform.

4 Proposed ICT Model

The ICT model for environment aware traffic management was developed as a layered model for making it to be flexible, so that it can adapt to any condition. The main advantage of the layered model is the ability to modify any layer without affecting the other layers. For example, if the communication between the sending units and the central processing center is implemented using SMS today, it can be easily replaced by another technology such as 5G or even 6G in the future without modifying any other component within the system. Also, it provides the application developers the freedom to choose the technology for implementing it on different platforms such as mobile devices, web etc. Figure 2 shows the proposed layered ICT model in this work.

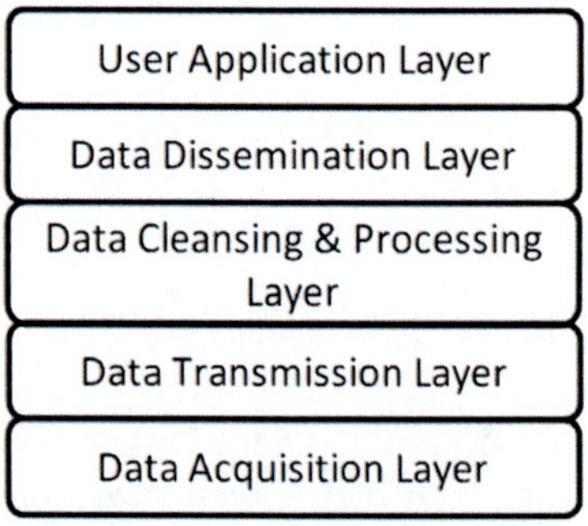

Fig. 2. Proposed layered ICT model

The implementation framework of the proposed model uses advanced ICT technologies available at present. The input data for the system includes both the current environment conditions at selected locations of a city and the information provided by motorists in terms of their current locations and intended destinations and other parameters of vehicles. The current environment conditions is collected using a IoT powered network of sensors. The vehicle information including the current location and intended destination, type, make, model and year of manufacture of the vehicle, and the type of fuel used are collected from the motorists' mobile devices. The data collected is transmitted to the central control implemented using cloud computing technology. The central control employs advanced data analytics application for predicting future environment conditions in terms of selected pollutants at the selected locations. Based on the current status and the predicted conditions, the system will generate the best route between the source and the destination for each vehicle and transmit it back to the mobile device using a set of parameters. At the receiver (mobile device) the user application will generate the route to the destination and present to the user as a driving direction map superimposed on the map of the city with instructions. Figure 3 shows the implementation framework of the proposed model.

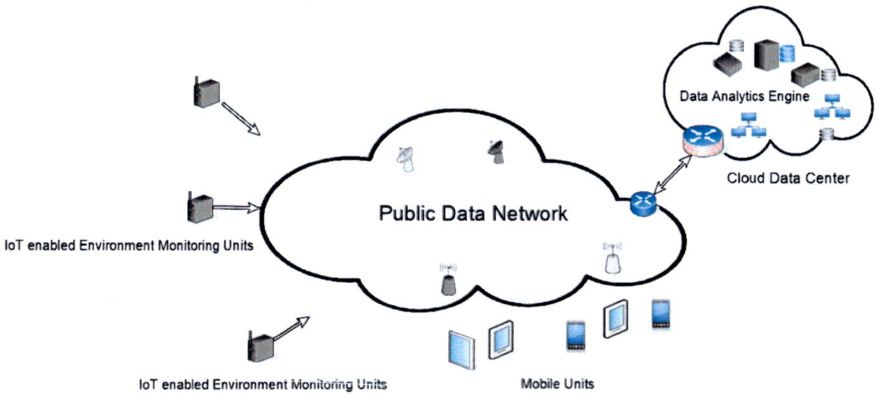

Fig. 3. Implementation framework of the proposed ICT model

5 Implementing and Testing the Model

The proposed model was implemented on the SUMO Traffic Simulator. SUMO is a full featured simulation suite with multiple traffic modeling utilities including a road network capable to read different source formats, demand generation and routing utilities from various input sources, a high performance simulation engine capable of simulating a single junctions or whole cities including a remote control.

5.1 Simulation Environment

The simulation environment was setup on the SUMO traffic simulator as follows:

A simple road network with one junction, four segments and five nodes was created as shown in Fig. 4. The nodes were named A, B, C, D and E and segments as A-B, B-C, B-D and B-E. In order to simplify the tests, all the road segments were defined as single lane roads that can carry only uniflow traffic starting from Node A and ending in either Node D, Node C or Node E. It is further assumed that a vehicle arriving at Junction B can take a left turn or right turn and can reach their destination through D or E and the same destination can be reached by the through traffic through Node C. Environment sensing unit was installed at Node C and the segments B-D and B-E are assumed to be far away from Node C and hence, the pollution created by vehicles traveling on those segments will not reach Node C.

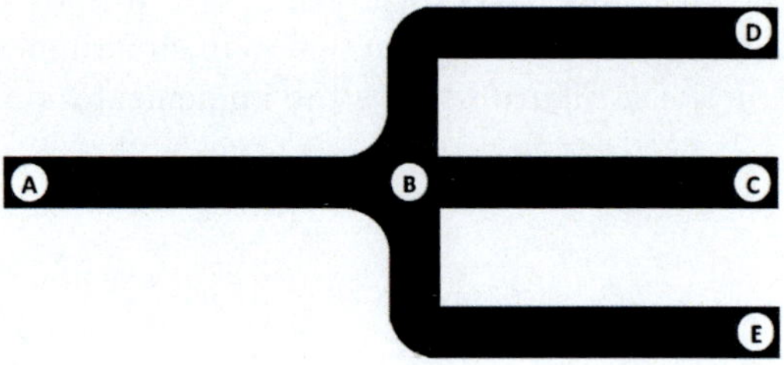

Fig. 4. Simple road network used in simulation

Only two types of vehicles were included in the tests as low polluting vehicles and high polluting vehicles and they are named LPVs and HPVs. It is further assumed that an LPV creates one unit of pollution and an HPV creates five units of pollution at Node C when it reaches it. The pollution decay is assumed to be uniform with one unit per unit time. The traffic generation was carried out using a uniform traffic generation model where only one vehicle (LPV or HPV) enters at Node A in a unit time and it will take any of the road segments at Junction B as instructed. The maximum pollution allowed at Node C is 80 units and when the pollution is expected to reach this level, vehicles will be diverted at Junction B.

5.2 Results and Discussion

In order to test various conditions three cases have been developed for testing as follows:

Case 1: Equal number of LPV and HPV; different types of vehicles were dispatched one after the other alternatively.
Case 2: More LPV than HPV; the vehicles were dispatched in the ratio of LPV:HPV = 5:1.
Case 3: More HPV than LPV; the vehicles were dispatched in the ratio of LPV:HPV = 1:5.

Figures 5(a), 5(b) and 5(c) shows the pollution generated at Node C for Case 1, Case 2 and Case 3 respectively.

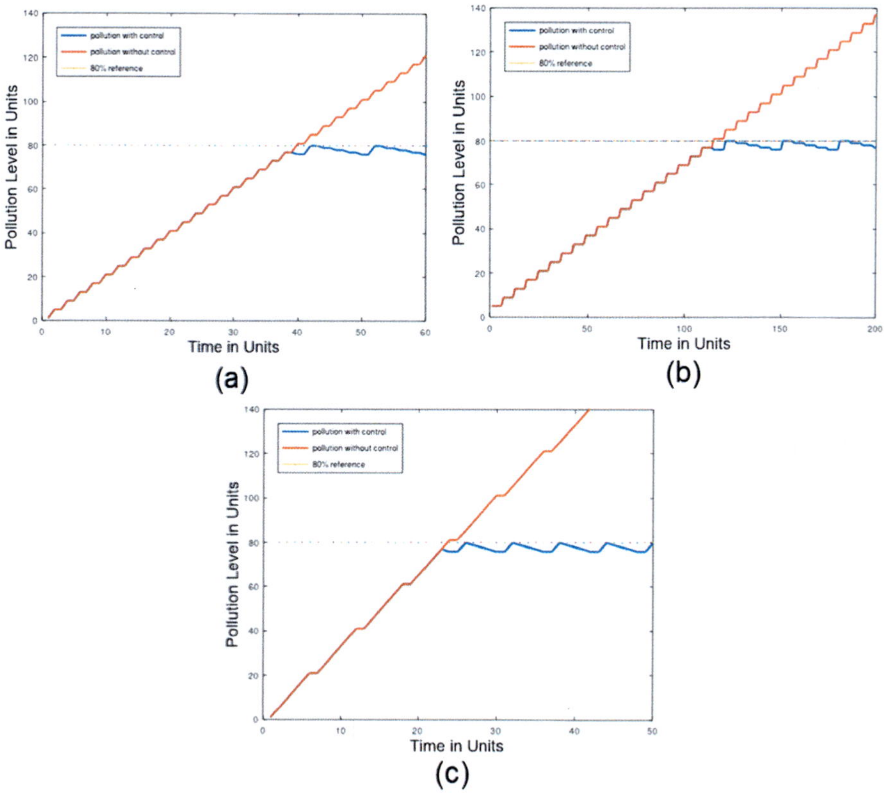

Fig. 5. Pollution at Node C under different input conditions

Prior to analyzing the results, it is to be noted that the graphs are plotted on different x-scales while keeping he y-scale fixed. This was done on purpose to magnify the effect of controlling vehicle entry into the location of interest. From Fig. 5, it can be observed that the pollution level at Node C continue to increase indefinitely, if the vehicle entry is not controlled. On the other hand, controlling vehicles help maintain the pollution level at the predetermined level at Node C. From Figs. 5(a) 5(b) and 5(c), it can also be observed that when more HPVs are allowed enter the road segment, the pollution level reaches the threshold level very fast compared to other cases.

When the threshold is about to be reached, the system starts diverting vehicles away from Node C. The pattern of vehicles diverted at Junction B is shown in Fig. 6. Figure 5 and Fig. 6 were plotted on the same horizontal (time) axis to help compare them with each other. The thickness of the bars is proportional to the number of vehicles diverted at Junction B. By comparing the three subplots of Fig. 6, it can be observed that the vehicle diversion starts sooner when more HPVs were allowed to enter the road segment leading to Node C. Also from Fig. 6, it becomes evident that only HPV vehicles get diverted once the pollution level at Node C reaches the threshold value. The reason for not diverting LPVs even after reaching the threshold level of pollution is that the pollution created by those vehicles practically disperses before the next vehicle arrives. Hence, it can be safely concluded that by introducing environment aware traffic management, it is possible to control the pollution at specific locations in cities by diverting HPVs away from those locations.

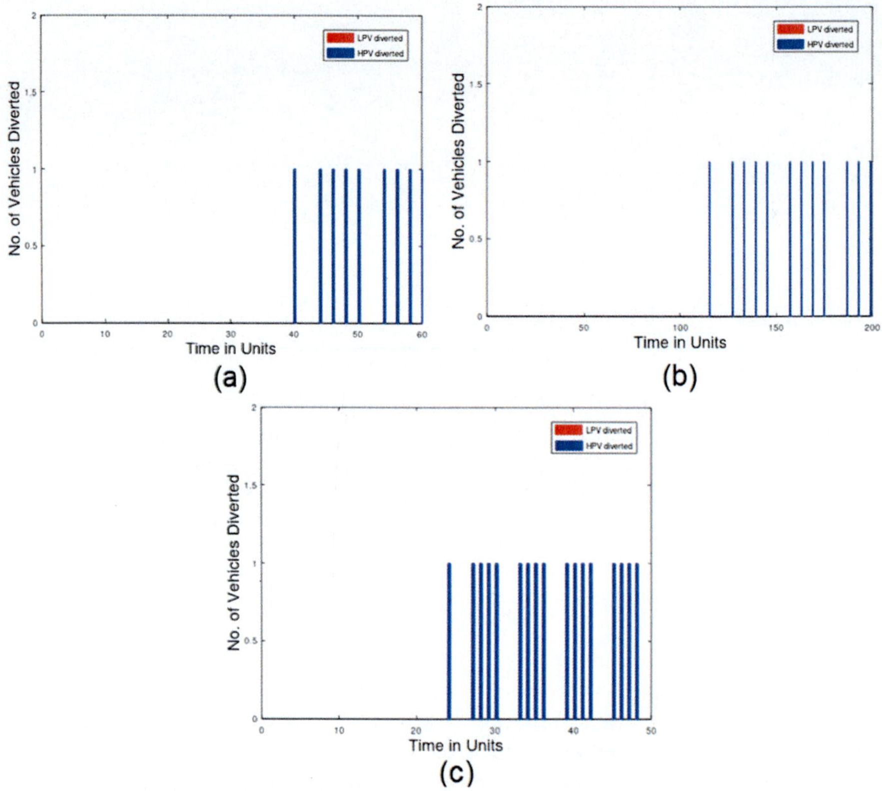

Fig. 6. Number and type of vehicles diverted at Junction B

6 Conclusions

In this paper, the authors proposed an IoT powered model for controlling vehicle induced pollution in cities. The proposed model has been tested under simulated conditions using the open source SUMO traffic simulator. From, the tests carried out as the proof of concept shows that the results are promising and will have real practical significance, if implemented properly. The tests carried out and the traffic and the pollution prediction models used in this work are very simple. Hence, the authors propose to extend this work in future using more realistic models for traffic generation, prediction and management using different types of vehicles with real parameter values as proposed in Sect. 2. Also, it is proposed to use more extensive data manipulation techniques including artificial intelligence for analyzing the environmental impact created by different types of vehicles to come up with more realistic recommendations.

Acknowledgments. This research was supported by the Accelerating Higher Education Expansion and Development (AHEAD) project of the Ministry of Higher Education of the Sri Lankan Government funded by the World Bank.

References

1. Akhter S, Ahsan MD, Quaderi SJS, Al-Forhad MA, Sumit SH, Rahman MR (2020) A SUMO based simulation framework for intelligent traffic management system. J Traffic Logistics Eng 8(1):1–5
2. de Souza A, Brennand C, Yokoyama R, Donato E, Madeira E, Villas L (2017) Traffic management systems: a classification, review, challenges, and future perspectives. Int J Distrib Sens Networks 13:1–14
3. Elmansouri O, Almhroog A, Badi I (2020) Urban transportation in Libya: an overview. Transp Res Interdisc Perspect 8:1–7
4. Firdhous MFM, Sudantha BH, Hussien NA (2021) A framework for IoT-enabled environment aware traffic management. Int J Electr Comput Eng (IJECE) 11(1):518–527
5. Huang Y, Wang L, Hou Y, Zhang W, Zhang Y (2018) A prototype IoT based wireless sensor network for traffic information monitoring. Int J Pavement Res Technol 11:146–152
6. Lalitha K, Pounambal M (2018) Density based traffic management using IoT. Int J Pure Appl Math 120:781–792
7. Makino H, Tamada K, Sakai K, Kamijo S (2018) Solutions for urban traffic issues by ITS technologies. IATSS Res 42:49–60
8. Malik S, Nafis M, Khan M (2018) IoT enabled traffic control model using Raspberry PI. Int J Adv Res Comput Sci 9:157–160
9. Manisalidis I, Stavropoulou E, Stavropoulos A, Bezirtzoglou E (2020) Environmental and health impacts of air pollution: a review. Front Public Health 8(14):1–13
10. Matharia N, Dave S (2018) Smart traffic management system using IoT. Int J Comput Eng Appl XII:1–4
11. Shaaban K (2019) Drivers' perceptions of smartphone applications for real-time route planning and distracted driving prevention. J Adv Transp 2019:1–10
12. Sowmya M, Anusha P, Murthy A, Jones J, Pinto L (2017) IoT based traffic management system. Int J Innov Res Comput Commun Eng 5:16408–16411

Cloud Security as a Service Using Data Loss Prevention: Challenges and Solution

Mohammad Equebal Hussain[1(✉)] and Rashid Hussain[2]

[1] Computer Science and Engineering, Suresh Gyan Vihar University, Mahal, Jagatpura, Jaipur, Rajasthan 302025, India
mdequebal.60508@mygyanvihar.com
[2] Suresh Gyan Vihar University, Mahal, Jagatpura, Jaipur, Rajasthan 302025, India
rashid.hussain@mygyanvihar.com

Abstract. Due to recent worldwide pandemics most of the organization allowed working from home. Employees are given access to sensitive information including company secure assets which need to be protected at any cost. This all happened in such a short span of time that company didn't get enough time to protect its assets due to lack of proper security software to protect data. In this paper we are focusing on a small subset of cloud security called data loss prevention (DLP). Data are broadly categorized into three states. Data at rest, data in motion or data in use. Files, directories and database tables are into DAR category, cut/copy/paste and file access comes under data in use whereas network packets, emails are in later category. These types of data may contain private information (PI) which needs to be protected in order to avoid misuse from an unauthorized user. In this paper we will discuss the challenges as well as propose a solution to achieve security in this area.

Keywords: Security as a service (SecAAS) · Intellectual property (IP) · Personally identifiable information (PII) · Personnel health information (PHI) · Health Insurance Portability and Accountability (HIPAA) and Health Information Technology for Economic and Clinical Health (HITECH) · Data loss prevention (DLP) · Data at rest (DAR) · Data in motion

1 Introduction

Covid19 which is the most global, secular, non-discriminatory and fair living being on this earth. It doesn't make any distinction among rich and poor, believers and non-believers, developed and underdeveloped. One thing is for sure that life before and after Covid19 will never be the same again [11]. This has put a temporary brake on fast pace running life to bring on grinding halt and a long pause. But, we have to emerge as victorious in this fight by focusing on the destination without missing beauty of the journey. We have to remember that every such transition opens new opportunities for us. This virus is spreading primarily due to close interaction between humans therefore best remedy recommended by specialist is nothing but social distancing. In coming future

R. Misra et al. (Eds.): ICIoTCT 2021, LNNS 340, pp. 98–106, 2022.
https://doi.org/10.1007/978-3-030-94507-7_10

this could be achieved by using automation, robotics, drone, AI based devices and IoT (Internet of Things). These all players are possibly the clients of cloud based control server which is used to manage and control those devices. These devices will generate huge data, needs high bandwidth, infinite storage, best networking support and the most important is security.

Millions of devices across the world are connected through internet and IoT machines. In coming years, these numbers will rise exponentially. Mobile phones will remotely control the IoT machines and possibly without physical touch i.e. a step forward towards social distancing. Working from home is not a new concept but post Covid19, companies will definitely analyze towards cost saving in office space, reduced transport expense, and increased employee performance. WFH will open new opportunities for "Desktop as a service" model using WFH tool kits.

In order to avoid touch systems, digital payment will be preferred over currency notes, also avoid touching biometric attendance, electrical switch, elevator buttons, doorknobs etc., new opportunities like voice based or facial recognition based attendance system, thermal camera to prevent sick employees to enter office etc. will be preferred which opens the new future for contactless technology post Covid19.

Virtual meetings and video conferencing are replacing International business travel. Lesson learnt from Covid19 is that medical staffs are at most risk treating virus positive patient. To protect health workers from direct contact, telemedicine contactless device with support of wireless sensor network (WSN) can be adopted to handle large number of patient without coming to hospital. Infrastructure required for development to support virtual telemedicine using wireless sensor network (vWSN) is another open door for opportunities.

Education sector is also not untouched from Covid19 waves. Institutions are closed. Online classes through video conferencing have already started. This opens new doors of opportunities for e-learning. It has many advantages like recording sessions, saving in commuting time etc. but due to huge pressure on cloud based storage and access to these services, security of personnel identifiable data is again at risk [6].

We mainly focused upon security of protecting private data because government may authorize agencies to collect the whereabouts using mobile phone location, use of debit/credit cards and similar personnel identifier for tracing of movement of a person which results in loosing right to privacy. The chance of leaking these data is maximized if proper security is not in place.

Threat could be from outside the organization such as virus attacks and unauthorized access. Threat could be from inside the organization such as sharing sensitive information using email. The outside threat could be prevented using firewall technique, encryption of content while storing data [4]. The DLP technique helps to protect organization's defined sensitive data from internal threats.

2 Background and History

Data is growing exponentially as we progress towards more technological environment. A data according to company policy is called sensitive if needs to be protected. Protection of sensitive and confidential data is the main goal of a company. Data leakage creates a negative impact and revenue loss for the company [5]. Security solutions like firewalls are required but not sufficient to protect data leak. Data leakage can be prevented using a dedicated solution called data loss prevention (DLP) system that protects sensitive data [8]. Organization's sensitive data are used across various services such as cloud and email; it is accessed through multiple devices like laptops and mobiles within or outside of the organization's network. In this situation it is very challenging and difficult to make sure that data is safe.

DLP secures sensitive data through automation by making sure that data is not stored, accessed or sent, where it is not suppose to send. Its action is tightly coupled with the sensitive information. For example, an email won't leave organizations network if it contains any sensitive data.

DLP handles access control, sensitivity level and various custom and predefined policies. Core feature of DLP includes email and file system monitoring, endpoint protection and network monitoring. Modern DLP also supports protecting data even if device is not within organization's network.

Types of data loss are mainly divided into three categories i.e. accidental loss, loss due to internal and external attacks. Accidental data loss could be because of employee unfamiliar in recognizing the sensitivity of documents and the company policies. In this paper an attempt is made to design and study DLP system for intellectual property (IP), personally identifiable information (PII), Personnel health information (PHI), Health Insurance Portability and Accountability (HIPAA, act 1966) and Health Information Technology for Economic and Clinical Health (HITECH, updated 2009).

Both HIPAA and HITECH are regulation and compliance standards to protect ePHI from intentional or unintentional exposure of sensitive information like Name, Phone, SSN, credit card, driver license, account number, Email, photograph, biometric information and address to an unauthorized party.

DLP solution allows organization to continuously monitor and control the movement of data. It is based on content analysis engine which is used to examine the content of files and emails in order to look for sensitive information, scans the documents before transfer. The actions could be logging the event, blocking email or warn end user who is sharing email or files.

The importance of DLP is illustrated in (Fig. 1) by Gartner's Hype Cycle for Storage and Data Protection Technologies, 2020 (ID: G00441602) security trend.

All diagrams except Fig. 1 are drawn using online drawing tool [18].

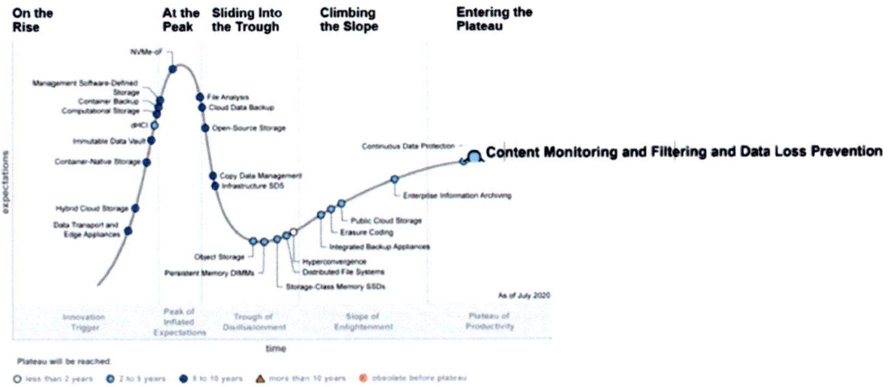

Fig. 1. Gartner hype cycle for storage and data protection 2020 (edited) [15]

3 DLP Technology

DLP system analyzes the data using two different ways. First method is content analysis which deals with actual data whereas the second method focuses on context analysis which deals with surroundings of data.

3.1 DLP Elements

Data loss prevention (DLP) as a security product or security as a service (SecAAS) assures that no sensitive information is sent outside the network. Using DLP, an administrator has full visibility and control over the data. In order to deliver functionality of DLP following elements are required [12].

- Register Data (data can be files, database with many records etc.), the registered data will be recorded as unique discrete hash value/fingerprint.
- Registered data (Data at rest or Data in motion) will be incorporated into policy. In case of data at rest, the inspection service will scan through each files looking for registered data by matching hash values. For data in motion, the inspection service will look for network traffic during transfer.
- Generate incidence if any protected content is detected. Depending upon the configured policy, file is blocked, allowed or retained.
- Predefined or custom file filters can be used to limit the files that are registered to meet the criteria.
- Data prioritization - not all data is equally critical
- Understanding data at risk - data in use is at highest risk when shared with partners and customers.
- Monitoring - to gain visibility.

3.2 States of Data

Different state of data helps understand to select appropriate security measure and encryption required to protect it. Data [1] can be in any of the three states: (a) Data at rest, (b) Data in motion and (c) Data in use.

Data at rest refers to those data which is stored on hard drive, cloud backup, database tables, USB devices etc. Data at rest is mostly inactive data, not actively being read and not currently being transmitted. Additional layer of securities like encryption, authentication and access control rules are required to protect this type of data.

Data in motion refers to those data [2, 3] which is currently travelling across network or present in computer's memory for ready to read, update and process. Data crossing over network should be encrypted using HTTPS, SSL, TLS, etc. to save from hackers. Data over wireless transmission, emails, and files transferred over FTP/SSH are some of the data in this category.

Data in use refers to data which is being processed, updated, appended or erased by applications, passively stored on hard drive. This data is visible through various end points during access. These data should be protected through strong user authentication, identity and profile management. Permission is granted through Non Disclosure agreement (NDA).

3.3 Securing Application in the Cloud

Simplified application development is offered by cloud computing using ready to use platform, infrastructure and software services. Cloud is a shared responsibility model therefore security concerns are majorly deciding factors while planning for cloud adoption, risk management and compliance. Effective control framework is required for continuous monitoring of incidence and threats are essential to move enterprise application over the cloud. Micro service architecture (Dockers and kubernetes) is a good fit for secure cloud application. Security is a joint responsibility between cloud provider and customer.

3.4 Secure Cloud Solution Key Components

Security is required to protect business, application, data and infrastructure in order to design, build, deploy and manage cloud based applications. Key components for secure cloud [7] applications are:

- Manage access – Managing user and customer access.
- Secure apps and data – identify and manage virtual machine vulnerabilities, encrypt data at rest (DAR) and data in motion. Auditing and monitoring of data activity.
- Gain visibility – collect and analyze real time logs, event and activity.

In order to provide end to end cloud security, unified solution is required for different types of deployment model including but not limited to public, private and hybrid model. Security components required are identity and access management, network security, application security, data security, monitoring solution, security policy, risk and compliance.

4 Software Architecture and Design

A DLP system is a policy-based mechanism for managing data discovery and classification on a user's computer called endpoint. Policies [8] are selectively distributed and imposed on each endpoint based on its characteristics. Not all information requires same level of security. Department of homeland security defines two types of PII, regular and sensitive. Business card falls under general PII whereas sensitive PII refers to information which if disclosed could result inconvenience to an individual. Therefore sensitive PII needs higher level of protection. Example social security number (SSN), license, passport number and biometric information is classified as sensitive PII. Developing a security strategy is important to protect the data. For a given endpoint, the applicable sets of policies are determined based on identity and other characteristics [9].

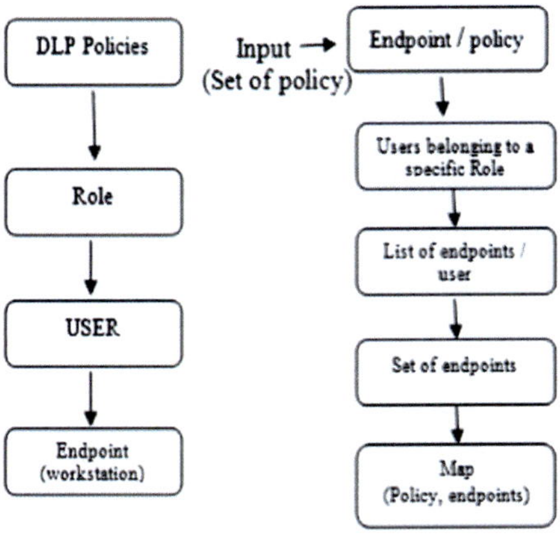

Fig. 2. Policy flow model

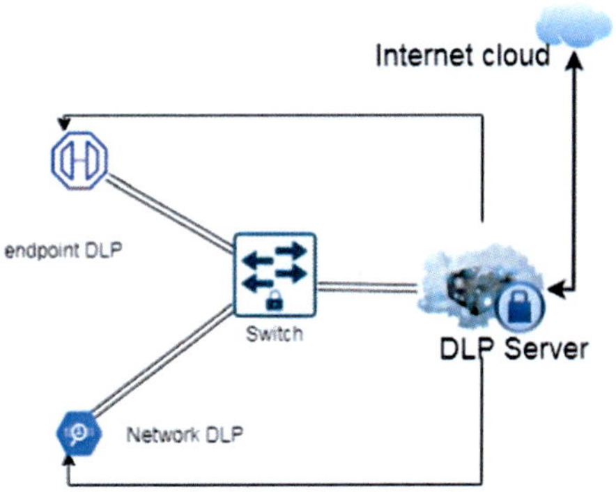

Fig. 3. DLP components

Figure 3 represents components common to any DLP system. Endpoint DLP keeps track of data at rest and data in use. It is directly installed on workstation. Data in motion (network traffic) is monitored through network DLP by placing between LAN and WAN as a proxy. Both network and endpoint DLP is managed by DLP server which is also responsible for policy deployment and logging the violations.

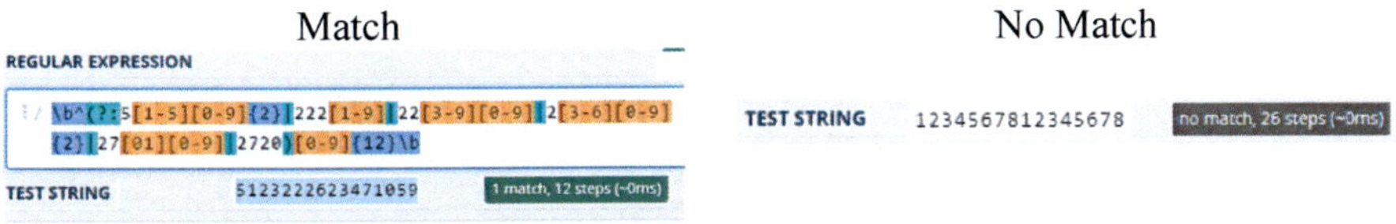

Fig. 4. Regex performance on matching vs. non matching pattern [13]

4.1 Algorithm

DLP engine is based on regular expression pattern matching [14] along with other indicators used to identity sensitive information. For example not all 16 digits is a valid credit card number. DLP is smart enough to distinguish between a valid credit card number and random 16 digits string (serial number, ticket number, part number etc.) using regular expression along with additional validation and check digits. Keyword such as “VISA”, “expiry date”, “CVV” within finite proximity is also an indicator of presence of credit card number in the document. To reduce false positive, DLP engine [12] uses proximity and confidence level in the evaluation process.

Figure 2 represents the flow diagram, how the policy is applied to a group of endpoints. This is applicable to either existing policy when it is modified or new policy. User or endpoints, who are member of role within policy target. One user may have ownership of more than one endpoint. Once the endpoint is identified then make a set of those endpoints. Once all the user of a particular role is identified then add the pair (policy, set of endpoints) to the output list [10].

The mentioned functionality may be implemented as a Restful API, web services via JSON or SOAP/XML interface.

In Fig. 4, we can see that regex engine took 12 steps to report matching a valid credit card number whereas 26 steps to report non-matching string.

4.2 DLP Policy Overview

Figure 5 shows that policies are the top level [16] objects which are converted into rules. The rule defines sensitive content. The rules are then either deployed to endpoint agent(s) or network DLP to protect data in different state which could be at rest, in motion or in use.

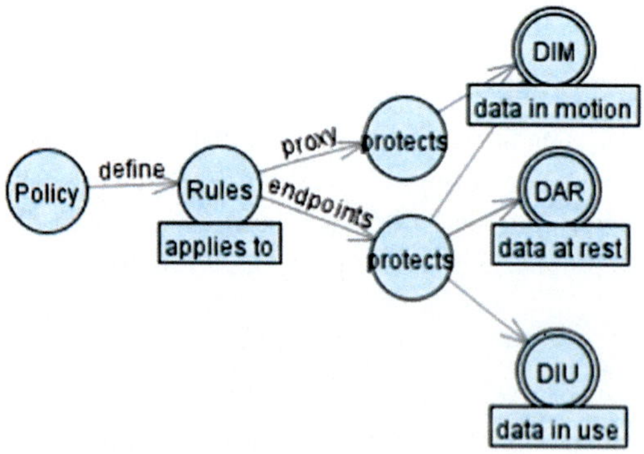

Fig. 5. DLP policy flow diagram

4.3 Proposed Design

Figure 6 represents a high level DLP design and its main components. In the proposed model, DLP policy is a combination of regular expression with terms or dictionary.

Terms and dictionary contains the list of confidential keywords corresponding to the policy. Multiword phrase can be further divided into single words called tokens before generating the policy using DLP engine.

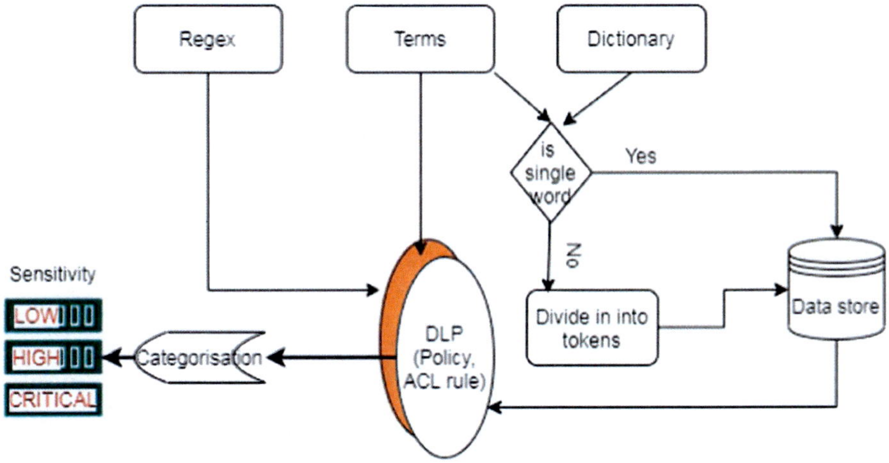

Fig. 6. DLP policy proposed design

4.4 Result

See Table 1.

Table 1. Result and performance analysis.

Description	Sample size (KB/MB)	Running time (MM:SS)
Normal as well as contextual Contains both Pattern (regex) and Terms Verified on virtual Email Security Appliance (Linux based) Model: C300V	100 KB	0 min 1.1 s
	200 KB	0 min 1.27 s
	300 KB	0 min 1.44 s
	1 MB	0 min 2.67 s
	5 MB	0 min 9.76 s
	10 MB	0 min 18.61 s

5 Conclusion and Future Work

The result shows that DLP policy is quite efficient and performs well when combined with regular expression, dictionary and terms. Maximum and minimum score can be defined by the policy. Total score is sum of individual scores of sensitive information matches within the document. Once the maximum score is reached, further scanning can be stopped to enhance the performance. Currently there is a limitation that only exact predefined keyword match will result a score [17]. This work can be further extended to apply machine learning technique and AI framework to match closely related keywords. This will further protect the confidential and sensitive information with minimal or no false positive. The other enhancement is to support optical character recognition (OCR).

References

1. Hart M, Manadhata P, Johnson R (2011) Text classification for data loss prevention. In: Fischer Hübner S, Hopper N (eds) PETS 2011, vol 6794. LNCS. Springer, Heidelberg, pp 18–37. https://doi.org/10.1007/978-3-642-22263-4_2
2. Mogull R (2009) Best practices for endpoint data loss prevention. s.l.: Securosis, L.L.C.
3. Mogull R Understanding and selecting a DLP solution: part 2, content awareness. https://securosis.com/blog/understanding-and-selecting-a-dlp-solution-part-2-contentawareness
4. Takebayashi T, et al (2010) Data loss prevention technologies. Fujitsu Sci Tech J 46:47–55
5. Faiz MF, Arshad J, Alazab M, Shalaginov A (2020) Predicting likelihood of legitimate data loss in email DLP. Future Gener Comput Syst 110:744–757
6. Noghondar ER, Marfurt K, Haemmerli B (2012) The human aspect in data leakage prevention in academia. In: Reimer H, Pohlmann N, Schneider W (eds) ISSE 2012 securing electronic business processes. Springer, Wiesbaden, pp 137–146. https://doi.org/10.1007/978-3-658-00333-3_14
7. Fusenig V, Sharma A (2012) Security architecture for cloud networking In: 2012 IEEE international conference on computing, networking and communications, cloud computing and networking symposium
8. Ashraf MA, Jamal H, Khan SA, Ahmed Z, Baig MI (2016) A heterogeneous service-oriented deep packet inspection and analysis framework for traffic-aware network management and security systems. Access IEEE 4:5918–5936
9. GitHub (April 2012) MyDLP. GitHub.com. https://github.com/mydlp/mydlp
10. Sonntag M, Oleshchuk VA (2012) Data loss prevention systems and their weaknesses. University of Agder
11. Ouellet MC, Gagnon E, Tourigny A, Lamontagne MÈ, Routhier F (2021) Impact of the COVID-19 pandemic on older adults: rapid review. JMIR 4(2):e26474
12. Tahboub R, Saleh Y (2014) Data leakage/loss prevention systems (DLP). In: IEEE conference, 17–19 January 2014. https://doi.org/10.1109/WCCAIS.2014.6916624
13. https://regex101.com/
14. Deaton S, Brownfield D, Kosta L, Zhu Z, Matthews SJ (2017) Real-time regex matching with apache spark. In: IEE conference, 12–14 September 2017
15. https://dellfocusaccounts.com/wp-content/uploads/2020/10/Hype-Cycle-for-Storage-and-Data-Protection_2020.pdf
16. Victor OW, et al (2016) Data loss prevention and challenges faced in their deployments. In: International conference on information and communication technology and its applications (ICTA)
17. Cheng L, Liu F, Yao D (2017) Enterprise data breach: causes, challenges, prevention, and future directions. Wiley Interdisc Rev Data Min Knowl Disc 7(5):e1211
18. http://draw.io/ online tool to draw the diagram

Wireless Sensor Network Based Distribution and Prediction of Water Consumption in Residential Houses Using ANN

Mohammad Faiz(✉) and A. K. Daniel

MMM University of Technology, Gorakhpur 273010, India

Abstract. Over the past few decades the amount of water used in the urban areas of various states in India has raised serious concerns. The average amount of water in the aquifers in the country has started to decrease with higher personal water consumption. The water crisis in the urban area needs the attention of the researchers to find efficient ways to overcome the problem of water distribution. The proposed model of prediction is developed by using ANN (Artificial Neural Network) for large communities of different types. Wireless Sensor Network-based monitoring and data collection scheme of water consumption are used in the proposed model for communities of different types such as household-1 with three members, household-2 with four members, and household-3 with eight members. The model provides the concept of a water flow control sensor in different applications such as showers to reduce water loss. The system enables the distributors to distribute optimally to communities. The water crisis in residential areas can solve by this model.

Keywords: Wireless Sensor Network · Water consumption · Artificial Neural Network

1 Introduction

Water supply is an enduring and serious concern in all parts of the country. There are many regions of Uttar Pradesh suffering from indoor water distribution problems not only during the dry season but also throughout the year. The water supply has become a serious concern, owing to the shortcoming of rains in recent years and the decline in the water level of the dams. According to a survey in 2019, 34.47% of revenue lost due to water pollution, spills, thefts, unauthorized consumption, and estimation errors. In India, it was 36.24%, an average of more than $8 billion. The United Nations (UNESCO, 2015) report, says that 70% of the world population would suffer water supply shortages if existing conditions of water supply or distribution are retained and a quarter of the population suffers a serious water scarcity [1]. The total water requirement in India by the year 2050 will become 1,422 BCM (billion cubic meters), which will be much more than the total utilizable average water resources [2]. Most municipal and urban water authorities draw water from sources where consumption exceeds preservation. Water is

R. Misra et al. (Eds.): ICIoTCT 2021, LNNS 340, pp. 107–116, 2022.
https://doi.org/10.1007/978-3-030-94507-7_11

used in manufacturing, our homes, recreation, sanitation industries and agriculture. Many industrial processes depend on water, and different offices and schools cannot run without proper sanitation. Sanitation is one of the major elements of water consumption in urban areas and needs to address for efficient water distribution. The lack of sanitation and hygiene systems to eliminate human waste from inhabited areas contributes to significant health issues, like bowel worms and dysentery (WHO, 2012). If society keeps drawing water at an inappropriate amount, we will experience a water crisis by 2025 in urban areas [3, 4]. There may be an easy downfall plunge in the water crisis and a state of serious problems. We need to observe the residential and individual water consumption on a small and weekly basis to understand and suggest ways of increasing water conservation and efficiency in Uttar Pradesh. In the last decade, a variety of researchers have conducted prediction and forecasting studies in a variety of ways. The production and consumption of water must be in balanced order and the total amount of water used must equal the total amount of water provided. To ensure this water balance between demand and supply, future water demand must be accurately forecasted [5]. Our study has focused on household water consumption in Uttar Pradesh because personal and household water usage in Uttar Pradesh is essential because it is the third-largest demanding area, after industrial and agricultural use [6].

2 Related Work

In the present time most of the researches are focused on hybrid approaches such as ANN is combined with Wavelet Transform (WT), ANN combined with Fuzzy logic because the combination of such forecasting techniques are outperform a classical single-method approach. Hybrid techniques are capable of extracting more information available for data prediction [8]. Peng, H et al. proposed a predictive model for water consumption in construction which is based on Genetic Algorithm-Backpropagation (GA-BP) performs significantly higher which is improved by Particle Swarm Optimization (PSO) using grey relational analysis [9]. Piasecki et al. have compared Multiple Linear Regression (MLR) and the Artificial Neural Network (ANN) and their research showed the ANN was better in demand prediction [10]. In practice, time-series forecasts have been made using traditional linear models such as Auto Regressive Moving Average (ARMA) and autoregressive (AR) [11]. These models are simple to implement, but they have limitations in terms of improving forecasting accuracy because they are based on linear functions. When applied to a complicated data set where the structure is uncertain, ANN models are extremely powerful. It was pointed out that the autoregressive integrate moving average (ARIMA) models are often used in time series analysis for decades and have been used in a wide range of scientific applications due to their performance [12]. Santos et al. have used the artificial neural network (ANN) and support vector machine (SVM) to predict São Paulo's urban water demand in Brazil [13]. The research has shown that the ANN system is capable of predicting accurately for a large volume of data with high performance and less error. The role of weather was also considered in their research. Currently, the use of artificial intelligence techniques for water consumption and demand prediction primarily consists of two aspects: a time series-based and a multi-parameter-based prediction model. In the earlier research, it has been shown that soft computing techniques such as ANNs can best address non-linear problems in the management of water resource requests. As a result, ANNs can identify the complex and nonlinear modeling between the variables that impact water demand prediction. BP-APP has several advantages such as simple concept, low calculation costs, and only few parameters that need to be adjusted in the training [14]. Murthy et al. proposed model using multiple linear regression (MLR) to predict the water demand. Their proposed model used ultrasonic sensor for data collection of usage. A web based app is used to take the input from the user. The system is designed for small scale in domestic usage. The overall efficiency of obtained from their proposed system was about 90% [15]. In Portugal, various studies related to the prediction of drinking water consumption has revealed that the multiple linear regression model based on past data is helpful and therefore can be coupled with some other prediction models to minimize errors but regression model are less capable of training compared to ANN [16]. Based on the previous analysis, a high precision model is to be developed to strengthen the BP neural network. Furthermore, most of the present research results on water consumption demand prediction have been applied in vast areas, such as agriculture, industry. To the best of the author's knowledge in this domain, no studies on water demand prediction projects for water consumption in the urban areas of Uttar Pradesh have been published. As a result, in the proposed work, a prediction model of water consumption and demand prediction is developed with Wireless Sensor Network using BP-ANN.

3 The Proposed Data Model Design

One of the most suitable solutions for efficiently monitoring and distributing water is WSN based water distribution system. Although the only parameter measured is the volume of water flow consumption and optimized for water consumption. The WSN based system is built and operates in low power [17] and low cost to create large WSNs, via a mesh network, capable of transmitting data to the cloud and enabling consumers to provide consumption information online. Figure 1 represents the proposed WSN based water distribution system.

The distribution area is divided into different sectors and clusters for the distribution of water for household applications. The system includes sensor nodes used to calculate household water usage through a flow sensor and transmit this data to the base station via a radio frequency module (RF). The monitoring area is represented as a zone that is divided into sectors/clusters [17]. The sensor nodes were placed on the water tanks and water outlets of the houses in every colony and sensor nodes were consistently monitoring the in-flow and out-flow of water in the house. The majority of household water application includes toilets, showers, dishwashers, washing machines, and faucets. We avoid external landscaping water for convenience. We had examined the residential and personal usage of water on a low scale and on daily basis to understand and suggest ways to improve water conservation and efficiency.

The data was collected in real-time using a flow sensor for indoor water consumption considering ten households of the Madan Mohan Malaviya University of Technology, campus in the city of Gorakhpur, Uttar Pradesh with different categories (Type-1 representing ad-hoc teaching faculties with three people, Type-2 representing regular faculties

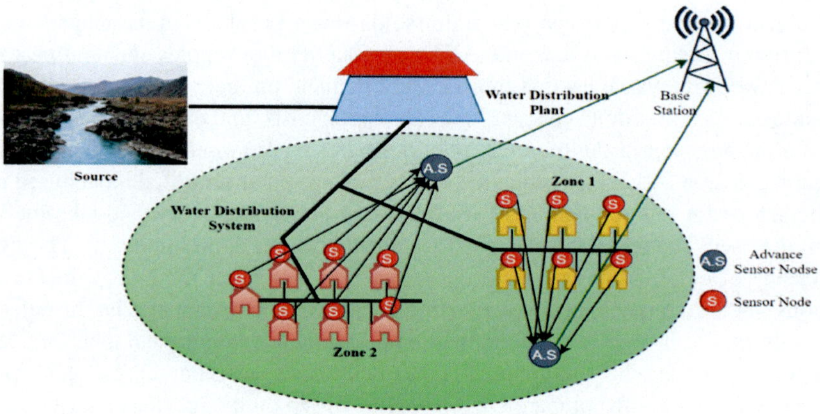

Fig. 1. WSN based water distribution system

Table 1. Weekly consumption of water in different households from different water outlets.

Sr.	A	B	C	D	E	F	G	H	Total
1	**Household1 (3 people)**	Mon	Tue	Wed	Thu	Fri	Sat	Sun	Week
2	Kitchen	9	7.0	21.4	11.7	5.9	9.7	4.7	69.4
3	Bathroom	4.2	4.0	5.0	4.8	4	4.5	3	29.5
4	Laundry	0	0	0	0	0	32	16	48
5	Toilets	8	12	10	12	4	14	6	66
6	Shower	23.7	50.4	41.5	54.8	43.0	32.6	62.2	308.8
7	Total	44.9	73.4	77.9	83.3	56.9	92.8	91.9	521
8	Total = 521 gallons								
9	**Household2 (4 people)**	Mon	Tue	Wed	Thu	Fri	Sat	Sun	Week
10	Kitchen	23.4	8	15.2	4.1	6.7	8.9	16.1	82.4
11	Bathroom	8.6	5.0	5.7	2.0	1.5	2.0	3.9	28.7
12	Laundry	0	0	0	0	0	30	40	70
13	Toilets	14.4	17.6	14.4	14.4	17.6	11.2	8	97.6
14	Shower	21.1	39	50.4	60.7	42.3	52.9	59	325.4
15	Total	67.5	69.6	85.7	81.2	68.1	105.0	107	584
16	Total = 584 gallons								
17	**Household3 (8 people)**	Mon	Tue	Wed	Thu	Fri	Sat	Sun	Week
18	Kitchen	10.5	31.7	5.8	18.2	21.4	4.8	11	103.4
19	Bathroom	10.3	6.6	3.5	6.8	4.4	3.6	10.3	45.5
20	Laundry	20	31.4	0	0	32.8	52.8	58	195
21	Toilets	33.3	37.3	31.2	33.4	32.9	41	9.4	248.5
22	Shower	56.2	90.3	96.5	123	92	110.7	132.7	701.4
23	Total = 1294 gallons	130.3	197.3	137	181.4	183.5	212.9	251.4	1294

with six people and Type-3 representing non-teaching staff with 8 people) on weekly basis.

Similarly, ten household data is also collected where each household has sensor nodes to collect the water outlet information. The data was initially captured as the total volume of water in gallons for faucets, showers, dishwashers, washing machines, and toilets. The sensors are used to collect data at each input point of application. Water consumption information can be accessed in the future via a Smartphone/desktop application and is available for cloud storage [18]. The collected data helps both customers and water suppliers to be accessed and tracked. Figure 2 represents the bar chart for indoor water consumption in ten different households.

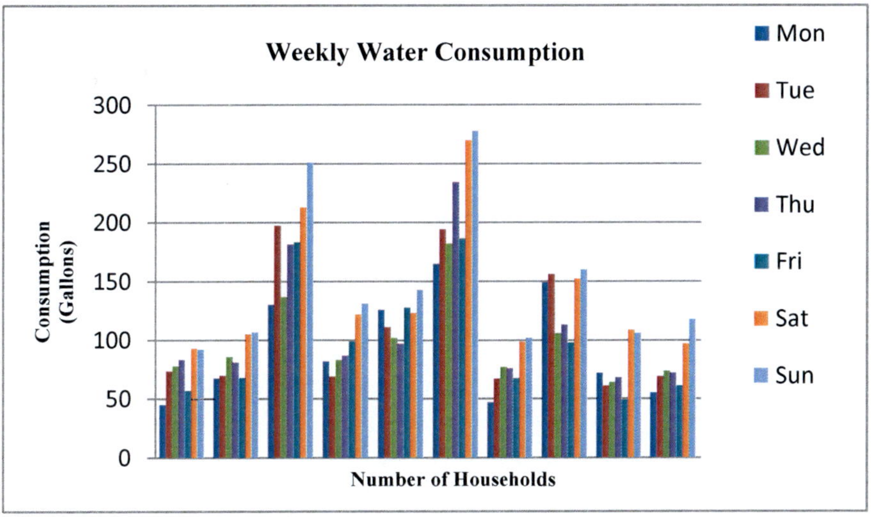

Fig. 2. Weekly consumption in different household

4 Proposed Model

The proposed Water Distribution Model (WDM) used a Backpropagation-Artificial Neural Network (BP-ANN) based on supervised learning to develop the model for household water consumption and demand prediction as an ANN is a system that represents a highly parallel distribution and data processing network similar to the human brain. Backpropagation refers to the backpropagation of error in the proposed BP-ANN prediction model [7]. The gradient of a loss function is computed through backpropagation with respect to different weights associated in the network. Proper tuning of the associated weights is performed to reduce error rates and make the model more reliable by enhancing its generalization. Apart from no. of input, no other parameters are needed to be tune in the model. The proposed BP-ANN includes an input layer that contains the input parameters of water sources (K, B, L, T, and S), a hidden layer, and an output layer that produces the output. The Backpropagation training algorithm is used in a feed-forward method that is interconnected by different layers. If an input set is provided to the input layer (input nodes), it propagates through the hidden layer until the output from the output layer (output nodes) is produced. The output is compared with the expected output where for each output node an error signal is computed. To modify the link weights, the error signals are then sent backward from the output layer (output node) to each node in the hidden layers until the error signal reached the expected need. The first layer of the model i.e. input layer contains P_j ($j = 1$ to 5) denoting values of considered five parameters used in weekly water consumption from different sources are Kitchen, Bathroom, Laundry, Toilet, and Shower. The hidden layer contains multiple neurons that are used to combine the weight matrix ($W_{i,j}$) and the bias vector (b_i). We can simply express that the computation of input parameter values using associated weight $W_{i,j}$ and bias b_i gave n_i value, which is subsequently transferred for substitution into f^1 called the transfer function. The transferred function applied here is called the log-sigmoid function to obtain the value of a_i^1. The a_i^1 is called the first hidden layer of the model. The mathematical model for the proposed WDM of water demand prediction is shown in Fig. 3.

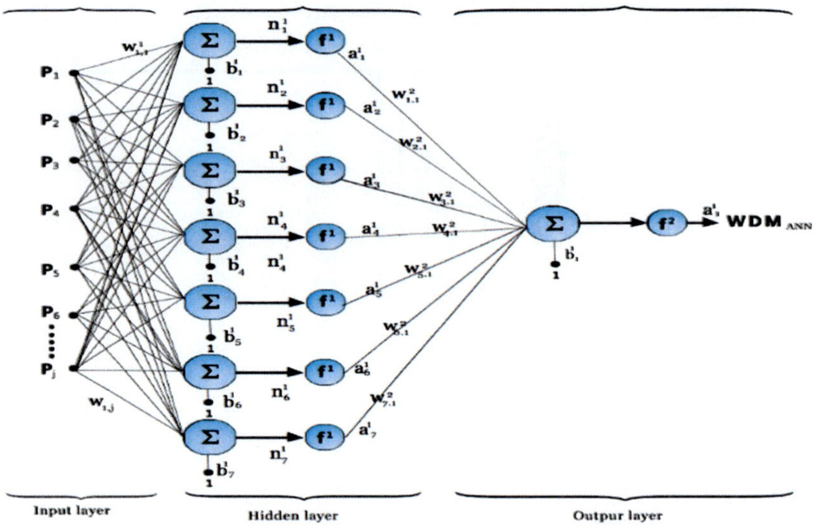

Fig. 3. WDM prediction model using BP-ANN

The computation steps for the Backpropagation algorithm are as follows:

$$a_i^1 = f^1(W_{(i,j)}^1 p_j + b_i^1) = f^1(n_i^1)$$
$$= logsin(W_{(i,j)}^1 p_j + b_i^1) \quad (1)$$

Where,

$$logsin(n) = \frac{1}{1 + e^{-n}} \text{(sigmoid function)}$$

The sigmoid function is used here because it exists between 0 to 1 and hence it is suitable for predicting the value based on probability in the proposed model.

i - The sequence number of neurons, j - The number of input ($p_1 = 1$ to 5).

The outcome of a^1 is connected to the next layer called the output layer, which contains a linear function f^2 to transfer the data further. The equation of the statement can be expressed as:

$$a^2 = f^2\left(W_{i,1}^2 a_i^1 + b_1^2\right) = f^2(n_1^2) \tag{2}$$

Algorithm: WDM prediction model Steps

Input: Input layer size P_i, Number of hidden layer n_i, Size of hidden layer h_s, Training data matrix X.

Output:

For each node

REPEAT

1. NO_CHANGES=FALSE

For each input parameter

Form a matrix for training (T), validation (V) and testing (T) of the neural network.

2. Initialize all weights randomly between [-1,1],set Bias=1, THRESHOLD=0.
3. Calculate the weighted input to the hidden layer by multiplying X by the weight $W_{i,j}$
4. Perform a forward pass to find the actual output (On).
5. Calculate the error Tn-On in the network, where Tn and On are target and output respectively.
6. If any error> THRESHOLD set NOCHANGES=TRUE
7. Do begin
8. Backpropagation to determine weight changes.
9. Update weights
10. End begin

UNTIL

11. NO_CHANGES
12. If the convergence is achieved then stop else repeat steps 1 to 6.

RETURN output

5 Simulation and Result

The simulation of the proposed WDM is performed in MATLAB. The collected data is observed as sample data used from Table 1. The inbuilt feed-forward function of MATLAB is used to create the neural network. In the training of ANN, the sigmoid function is used to map to real values. The simulation process subsequently performs adjustment of associated weights and bias until the convergence is achieved. The water consumption of three households such as households-1 households-2 households-3 along with members as 3, 4 and 8 respectively taken for prediction.

Figure 4(a)–(d) are representing the neural network training and testing with the target data set.

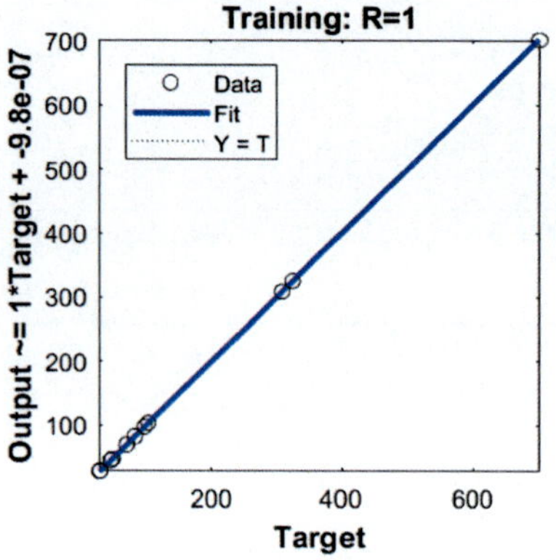

(a) Training of the model with past dataset

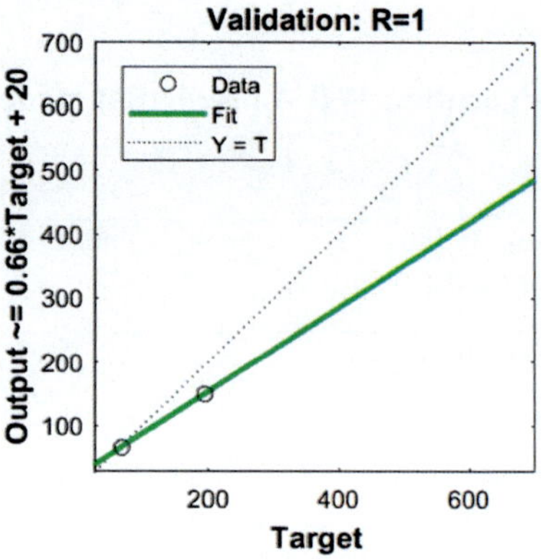

(b) Validation of the model Target data set

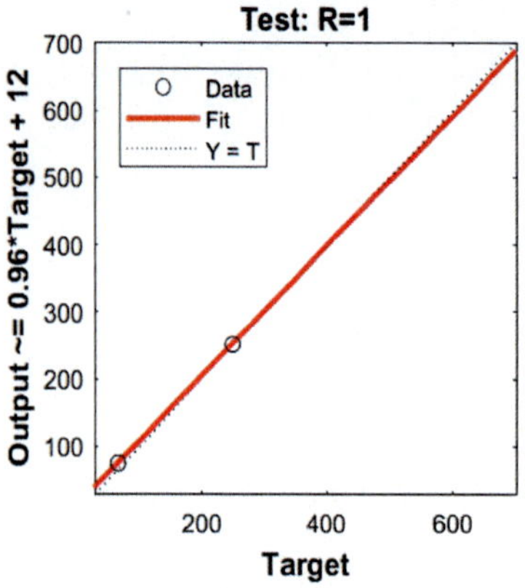

(c) Testing of the model

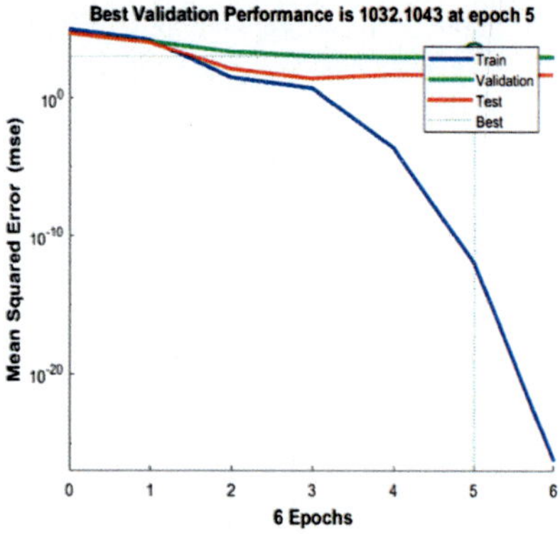

(d) Performance validation of the model

Fig. 4. (a) Training of the model with past dataset (b) Validation of the model target data set (c) Testing of the model (d) Performance validation of the model

The simulation process subsequently performs adjustment of associated weights and bias until the convergence is achieved. The water consumption of three households such as households-1 households-2 households-3 along with members 3, 4 and 8 respectively for prediction. The simulation is performed with a minimum gradient value of 10^{-6}. Figure 5 represents the comparison of actual consumption of water with regression model and proposed BP-ANN model.

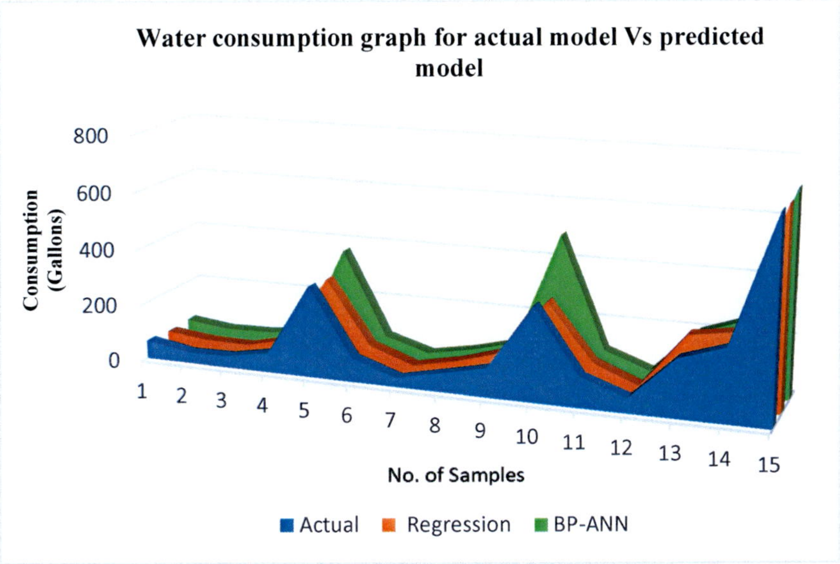

Fig. 5. Expected vs predicted values after training

6 Conclusion

The proposed model predicts the consumption of water for large communities of different types such as household-1 household-2 and household-3 having different amplitudes of people. The model suggested that using a flow control sensor for showers can reduce water consumption and it will be around 50% of household items. The model provides a correlation analysis function of 97.8% which indicates high similarity between actual and predicted data. The system enables the distributors to distribute water optimally to communities. The water crisis in residential areas can solve by this model.

References

1. UN, WWAP (United Nations World Water Assessment Programme) (2015) The United Nations world water development report 2015: water for a sustainable world. UNESCO, Paris, p 2015
2. Tiwari R, Nayak S (2013) Drinking water and sanitation in Uttar Pradesh: a regional analysis. J Rural Dev 32(1):61–74
3. World Health Organization (2012) Global analysis and assessment of sanitation and drinking water. Accessed 17 Nov 2012
4. Aral MM, Guan J, Maslia ML (2010) Optimal design of sensor placement in water distribution networks. J Water Resour Plan Manag 136(1):5–18. https://doi.org/10.1061/(asce)wr.1943-5452.0000001
5. Maroli AA, Narwane VS, Raut RD, Narkhede BE (2020) Framework for the implementation of an Internet of Things (IoT)-based water distribution and management system. Clean Technol Environ Policy 23(1):271–283. https://doi.org/10.1007/s10098-020-01975-z

6. Rondinel-Oviedo DR, Sarmiento-Pastor JM (2020) Water: consumption, usage patterns, and residential infrastructure. A comparative analysis of three regions in the Lima metropolitan area. Water Int 45(7–8):824–846. https://doi.org/10.1080/02508060.2020.1830360
7. de Menezes PL, de Azevedo CAV, Eyng E, Neto JD, de Lima VLA (2015) Artificial neural network model for simulation of water distribution in sprinkle irrigation. Rev Bras Eng Agric e Ambient 19(9):817–822. https://doi.org/10.1590/1807-1929/agriambi.v19n9p817-822
8. Piasecki A, Jurasz J, Kaźmierczak B (2018) Forecasting daily water consumption: a case study in Torun Poland. Periodica Polytechnica Civil Eng 62(3):818–824
9. Peng H, Wu H, Wang J (2020) Research on the prediction of the water demand of construction engineering based on the BP neural network. Adv Civil Eng 2020
10. Piasecki A, Jurasz J, Kazmierczak B (2018) Forecasting daily water consumption: a case study in town, Poland. Periodica Polytechnica-Civil Eng 62(3):818–824
11. Zhang W, Yang Q, Kumar M, Mao Y (2018) Application of improved least squares support vector machine in the forecast of daily water consumption. Wireless Pers Commun 102(4):3589–3602
12. De Souza Groppo G, Costa MA, Libânio M (2019) Predicting water demand: a review of the methods employed and future possibilities. Water Sci Technol Water Supply 19(8):2179–2198. https://doi.org/10.2166/ws.2019.122
13. does Santos CC, Pereira AJ (2014) Water demand forecasting model for the metropolitan area of São Paulo, Brazil. Water Resour Manage 28(13):4401–4414
14. Popa AS, O'Toole C, Munoz J, Cassidy S, Tubbs D, Ershaghi I (2017) A neural network approach for modeling water distribution system. In: SPE western regional meeting proceedings, April 2017, pp 990–1004. https://doi.org/10.2118/185678-ms
15. Chandrashekar Murthy BN, Balachandra HN, Sanjay Nayak K, Chakradhar Reddy C (2020) Prediction of water demand for domestic purpose using multiple linear regression. In: Smys S, Iliyasu AM, Bestak R, Shi F (eds) New trends in computational vision and bio-inspired computing. Springer, Cham, pp 811–817. https://doi.org/10.1007/978-3-030-41862-5_81
16. Ismail Z, Jamaluddin FA (2008) A backpropagation method for forecasting electricity load demand. J Appl Sci 8(13):2428–2434
17. Narayan V, Daniel AK (2021) A novel approach for cluster head selection using trust function in WSN. Scalable Comput Pract Exp 22(1):1–13. https://doi.org/10.12694/scpe.v22i1.1808
18. Faiz M, Daniel AK (October 2020) Fuzzy cloud ranking model based on QoS and trust. In: 2020 fourth international conference on I-SMAC (IoT in social, mobile, analytics and cloud) (I-SMAC). IEEE, pp 1051–1057

An Approach for Energy-Efficient Lifetime Maximized Protocol for Wireless Sensor Networks

Namrata Mahakalkar(✉) and Mohd. Atique

Department of Computer Science and Engineering, Sant Gadge Baba Amravati University, Amravati, India

namratamahakalkar@gmail.com, mohd.atique@gmail.com

Abstract. In order to optimise the network life of the wireless sensor networks (WSNs), the data transmission routes are chosen in a manner that minimises the total energy used along the way. Sensor nodes are organised into clusters to enable high scalability and better data aggregation. Clusters create WSNs that are hierarchical and capable of using limited sensor node resources, extending network life. An improved method for cluster heading selection is presented to increase the performance of traditional cluster head selection techniques. The DERDWSN and LBRWSN protocol is the hybrid version. In the study presented, the virtual idea is utilised for the creation of cluster head support which is useful for the duration of the network and for communication between the cluster head and the base station with efficient energy. The suggested procedure is carried out using the NS2 platform, examined and assessed in several simulated settings. The results of the simulation show that the protocol proposed is applicable and feasible and surpasses the performance of current algorithms.

Keywords: Wireless Sensor Network · Energy · Routing protocol · Network lifetime · Cluster · NS2

1 Introduction

The network of wireless sensors may be established by combining the number of wireless sensor nodes, which are a micro-electronic device fitted with a restricted power supply. [1] In certain application situations, it may be difficult to refill power supplies such that the lifespan of the sensor node relies heavily on the battery life. Each node performs the dual function of the data producer and data router for various applications, such a multi-hop ad hoc sensor network. Failure of nodes may lead to substantial network topology changes and may include re-routing of traffic and network reorganisation. Therefore, power management and power saving are of extra significance. Concentrating on the creation of power algorithms and protocols for sensor networks. Strom consumption has been a significant design issue for previous mobile and ad-hoc networks [2, 3] but this is not the main consideration, since the facilities are provided to enable users to replace power supplies when required, while focus is on QoS providing rather than power

R. Misra et al. (Eds.): ICIoTCT 2021, LNNS 340, pp. 117–128, 2022.
https://doi.org/10.1007/978-3-030-94507-7_12

efficient. Power efficiency is, nevertheless, an essential performance measure in sensor networks that directly affect network longevity. Application customised protocols may be created by trading other performance parameters properly, such as latency and power efficiency throughput. The primary job of a sensor node in a sensor field is the detection of events, fast local processing and data transmission. [4] Minimizing the energy used by the system is a crucial aspect of every wireless sensor node. The radio subsystem usually needs the most power. It is thus preferable to transmit data only when necessary via the radio network. This data collecting approach based on an event sensor needs an algorithm to be put into the node to decide whether data may be sent on the basis of the detected event. In addition, the power used by the sensor itself must be reduced to a minimum. The equipment should thus be built to enable the microprocessor to regulate the power of the radio, sensor and sensor signals.

1.1 Ad Hoc Protocols

Ad Hoc Network is a collection of two or more nodes or terminals that interact with each other without the help of any centralised administrator and also of wLAN nodes which may dynamically create a network for the exchange of information without the use of existing fixed network infrastructure. The ad hoc network should always be able to adapt to changing networks of this kind and mobile ad hoc network types.

1.1.1 "Destination-Sequenced Distance-Vector Routing (DSDV)"

The sequenced distance routing system is based on conventional Bellman-Ford routing algorithms and is based on the routing table. In an operation each node must be saved with a routing table that records all the possible connections with the node, and a distance such as hop number, and a sequence number in the routing table for each record which will be used to determine whether there is an old path to avoid routing table generation".

1.1.2 "Global State Routing (GSR)"

Global State routing is almost identical to DSDV, since it does have the concept of link state routing but progresses by reducing the inundation of routing packets. Each node has a neighbouring list, a topology table, a next hop table and a distance table in this method.

- The neighbouring node list lists its neighbouring nodes (all nodes that may be heard).
- The connection status information is kept in the topology table for each destination.
- Together with the information time stamp.
- The next hop table contains the next hop to which packets must be sent for each destination.
- The distance database includes the shortest distance to each target node.

1.1.3 "Ad Hoc On-Demand Distance Vector Routing (AODV)"

Ad hoc on-demand distance vector routing utilising the notion of a distance vector and in many different methods AODV does not keep a routing table but only communicates to the routing table approach if a node needs to contact another node on request. When a node wishes to transmit data to a new node on the Internet, the first node to submit a Route Request (RREQ) packet, RREQ, will record that a source is provided to determine the destination node. RREQ is a kind of flooding in transfer mode, destination until it is received, naturally the node cannot be processed just once on the same RREQ to prevent the creation of routing loops. In principle, all nodes between the source and the RREQ destination pass a temporary record on the final RREQ hop through Information path, if the RREQ destination is received from various locations, select the quickest way and send the route reply direction to the source (RREP). As the RREP for passing the nodes on this path is a registering of the necessary information, this segment of path from source to destination has been set up after the RREP has been delivered to the RREQ the source, and the source may then utilise this route to transmit packets to destination.

1.1.4 "Dynamic Source Routing (DSR)"

The use of the notion of source routing, the routing information that is stored directly within a packet and in order to be in the MANET environment, requires the use of such a specific DSR only when the route is necessary to find out a way on demand. Route Discovery with AODV is similar, but also broadcast to send a Route Request from a source client, it is, Route Request after one hop, this hop of the ID is recorded in Route Request a Route Record, the way in which the Route Request reaches a destination, all the nodes in the information route, a destination in many chosen route applications with the best patches".

"DSR Modifications, Extensions

i) Intermediate nodes may send route replies in case they already know a route
   - Problem: stale route caches
ii) Promiscuous operation of radio devices – nodes can learn about topology by listening to control messages
iii) Random delays for generating route replies
   - Many nodes might know an answer – reply storms
   - NOT necessary for medium access – MAC should take care of it
iv) Salvaging/local repair
   - When an error is detected, usually sender times out and constructs entire route anew
   - Instead: try to locally change the source-designated route
v) Cache management mechanisms
   - To remove stale cache entries quick Fixed or adaptive lifetime, cache removal messages".

2 Literature

In order to optimise the network life of the wireless sensor networks (WSNs), the data transmission routes are chosen in a manner that minimises the total energy used along the way. Sensor nodes are frequently organised into discrete, non-overlapping subsets termed clusters, to enable high scalableness and higher data aggregation. Clusters create WSNs that are hierarchical and capable of using limited sensor node resources, extending network life. The aim is to compare various cluster head selection methods described in the WSN literature. One of the most significant problems in WSNs is to develop an energy-efficient routing mechanism to improve network life because of the restricted network node energy capacity. In addition, hot spots in WSNs develop as high traffic locations [5]. Nodes in such locations rapidly deplete energy resources and lead to network services being disconnected. Recently, cluster-based WSN routing algorithms have gone up, demand and energy efficiency have become more selective. A cluster head (CH) represents and gathers data values from all nodes of the cluster [6]. The CH between all nodes is to be utilised and the cluster size should be carefully established in various sections of the WSN to balance energy consumption and traffic load in the network.

2.1 LEACH Algorithm

Using the distributed method, LEACH creates clusters, in which nodes take independent choices without central supervision. All nodes may become CHs to balance the energy each sensor node consumes. At first a node chooses to be a CH with a "p" probability and broadcasts its choice. More specifically, each CH will be broadcasting a message of advertising to the other nodes after its election and each (non-CH) node will decide its cluster by selecting the CH, which can be reached by using the least communication energy (depending on each CH-message signal strength).

The function of a CH is cycled regularly amongst the cluster nodes for load balancing. The rotation occurs when each node selects a random integer "T" of 0 to 1. For the current rotation cycle a node will become a CH if the number is fewer than the threshold:

Where,

"p is the desired percentage of CH nodes in the sensor population
r is the current round number
G is the set of nodes that have not been CHs in the last $1/p$ rounds".

2.2 "Energy Efficient Hierarchical Clustering

A new major probabilistic clustering method was presented previously. (Powerful Hierarchical Clustering—EEHC). The primary goal of the method was to overcome the weaknesses of one-hop random selection algorithms like LEACH by expanding the cluster architecture to several hops [7]. It is a distributed, hierarchical clustering method designed to maximise network life. First of all, every sensor node is chosen as a CH with "p" probability and its choice of the adjacent nodes within its communication range is announced. The CHs above are now referred to as "volunteer" CHs. Next, the elections should be sent to all nodes located within the "k"-hop distance from a "volunteer" CH, directly or intermediately. Any node receiving such a CH election message, which is not a CH, will thus become a member of the nearest cluster.

2.3 Hybrid Energy-Efficient Distributed Clustering

HEED is another enhanced and popular energy-efficient technique (Hybrid Energy-Efficient Distributed Clustering). HEED is a hierarchical, distributed clustering system in which the communication pattern for a single-hop is maintained inside each cluster, while multi-hop communication between CHs and the BS is permitted [8]. The CH nodes are based on two fundamental factors, residual energy and communication costs within clusters". The residual energy of each node is utilised to select the initial CH set probabilistically. On the other hand, intra-cluster communication costs mirror the neighbor's node or node proximity and are utilised by the nodes to decide whether or not to join a cluster. Thus, unlike LEACH, the CH nodes in HEED are not randomly chosen. Only sensors with substantial residual energy should become CH nodes.

2.4 ANCAEE Algorithm

"In order for a node to become cluster head in a cluster the following assumptions were made.

1) All the nodes have the same initial energy.
2) There are S nodes in the sensor field.
3) The number of clusters is K.

Based on the above assumptions, the average number of sensor nodes in each cluster is M where.

After M rounds, each of the nodes must have been a cluster head (CH) once.

We assigned each node a unique identifier i, Mi for all 0, 1, 2, 3, 4, S-1........

Variable i is used to test whether it is the turn of a node to become a CH.

Originally, all nodes are the same, *i.e.* there is no CHs in each cluster, $j = 0$ where j is CHs counter.

A node q is chosen from all nodes and continues to do the following steps:

First, q increases by 1 and verify whether I am equal, if yes, the node is chosen for the round as the CH and broadcasts its new position to all the nodes in the cluster. Alternatively, if I'm strange, that's not a CH for that round, wait for the next round and get advertising message from the new CH.

For the new CH to transmit for that round, a present value (threshold value) is established. Once the value is achieved, j is increased by 1 and the selection process of new CH starts. It checks if two criteria are met. That in the last 1 pp rounds a sensor node has not become the cluster leader, the node's residual energy exceeds the mean energy of all sensor nodes inside the clustering. The chance that a node will become a new cluster head is thus shown where the rest of the energy is averaging the energy of all nodes in the Eavg cluster in node I It goes on till j = K. When j = K, the algorithm ends". The new CHs gather, consolidate, and send sensed data from member nodes onto the next cluster head or base station.

2.5 "LEACH-DC Routing Protocol"

In the initialization of the network, LEACH-DC utilises LEACH-C architecture in [9]. In turn, the nodes drive to a sink. The sink estimates the distance from the centre of the region by each node and transmits it to every node. Add the current energy Ei-current and the original energy Ei-total for the node I to the threshold calculation for choosing the cluster head for a single energy ratio. For a node that consumes additional energy, T(n) values and the probability of the appropriate cluster head are reduced. In contrast, for a node that consumes less energy, the values of T(n) and the likelihood of an appropriate cluster head rise.

We have examined various routing algorithms and compared them to the LEACH protocol. LEACH and its advanced protocols have been published to date in WSN literature and have provided a comparison of some progress in the LEACH protocol [10]. Some energy-efficient methods have been discovered to improve network life and also waste energy in routing. Every attempt has been made to offer a comprehensive and precise state-of-the-art review of energy efficient clustering algorithms along with LEACH and its sophisticated WSN protocols.

3 Proposed Work

"Each cluster has a lead, which is also called the cluster head (CH), in the hierarchical network structure and typically performs the specific duties of fusion and aggregation as members and various common sensor nodes (SN). The process of clustering ultimately leads to a two-tiered hierarchy in which the CH nodes are at the highest and the cluster nodes at the lower level. Periodically, the sensor nodes send their data to the appropriate CH nodes. Since CH nodes transmit data to distances greater than common (member) nodes on a continuous basis, they spend higher energy rates. A typical method for balancing energy usage across all network nodes is the re-election of new CHs regularly rotating the CH role amongst all nodes in each cluster over time".

In DERDWSN model at the time of data transmission, after initialization of network parameters, source and destination node of the network is to be considered. Route between the source and destination is calculated with Euclidean's distance algorithm. Effective Minimum distance route leads to Communication and packet transfer between different nodes without forming cluster by using DSR protocol and Euclidean's distance algorithm (DSR based randomly deployed WSN).

In LBRWSN model Communication and packet transfer between different nodes as same as LEACH protocol by selecting cluster head and using sink node (LEACH based randomly deployed WSN) (**LBRWSN**). Performance Analysis of both the Technique in done using the simulation parameters in Table 1.

Simulation Parameters

Table 1 describes the simulation parameter used for simulation. Simulation scenario consists of nodes arranged in star topology. Each node is equipped with Omni directional antenna which can transmit signal in all directions. Throughput is 98.52% because the clusters are small so the inter-node distance is small and packet loss is nearly zero.

"Performance of the network is measure in terms of

Table 1. Initialising parameter for network

Sr. no.	Parameter	Values
1	Simulator	NS-2.34
2	Channel type	Channel/Wireless Channel
3	Radio-propogation model	Propagation/TwoRayGround
4	Network interface type	Phy/WirelessPhy
5	MAC type	Mac/802.11
6	Interface queue type	Queue/Drop Tail
7	Antenna model	Antenna/Omni Antenna
8	Max packet in interface queue type	50
9	Routing protocol	DSR
10	Dimension of the topography	1500 * 1500
11	Simulation time	40 s
12	Initial energy	100 J

1. Energy Consumption: This paper aims to lessen the energy consumption for the network. The average energy spends by both the technique measure and compare them, which technique requires less energy will be the useful technique in wireless sensor network. Initial energy taken for every node is 100.
2. Packet Delivery Ration (PDR): The ratio of the number of delivered per data packet to the destination. This illustrates the level of delivered data to the destination.
   $\sum$ Number of packet receive / $\sum$ Number of packet send
3. Throughput of System: The throughput is usually measured in bits per second (bit/s or bps), and sometimes in data packets per second or data packets per time slot. The system throughput or aggregate throughput is the sum of the data rates that are delivered to all terminals in a network.
4. Delays: The delay is the average time taken by a data packet to arrive in the destination. It also includes the delay caused by route discovery process and the queue in data packet transmission. Only the data packets that successfully delivered to destinations that counted.
   $\sum$ (arrive time – send time) / $\sum$ Number of connections".

Comparison will be done between different algorithms techniques based on these parameters & and results will be compared.

4 Simulation Results and Discussion

The simulation environment shown in Fig. 3 which contains total 49 plus nodes. It consist of a base station (BS) located at the center with coordinate 500 m × 500 m. The nodes are arranged in a star topology. Although the nodes distribution considered for simulation is in a line, but it is not necessary to have them in line. They can be distributed randomly. "It will evaluate the result for average energy consumption, packet drop, packet Delivery ratio, and throughput".

Results of DSR Protocol/Eucledean Distance Based Randomly Deployed WSN (DERDWSN)
"Here we deployed 49 nodes. In topology discovery process the simulation start searching for the source (transmitting) node and destination (receiving) node" (Figs. 1 and 2).

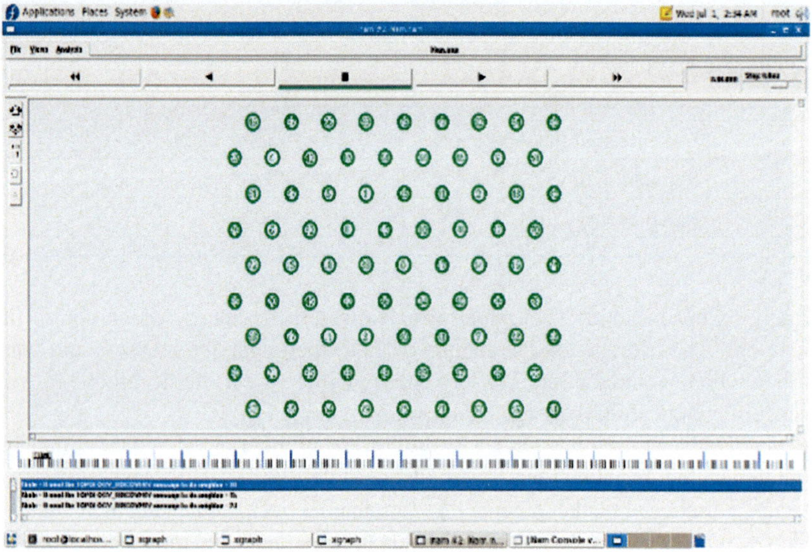

Fig. 1. NAM file1 of DERDWSN

Here in Fig. 3 we deployed time (x axis) vs throughput (y axis) of DERDWSN QoS parameter through which we can observe the throughput parameter on mentioned protocol.

In Fig. 4 we deployed time (x axis) vs PDR (y axis) of DERDWSN QoS parameter through which we can observe the packet delivery ratio parameter on mentioned protocol.

In Fig. 5 we deployed time (x axis) vs Energy (y axis) of DERDWSN QoS parameter through which we can observe the throughput parameter on mentioned protocol.

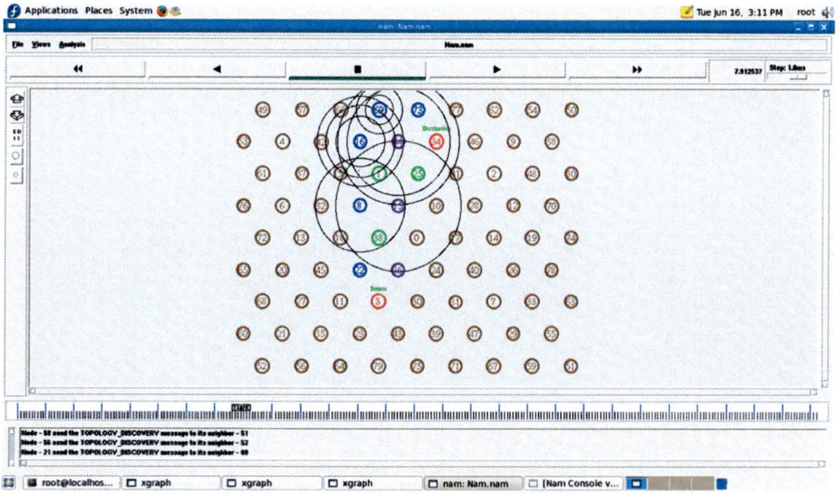

Fig. 2. NAM file2 of DERDWSN

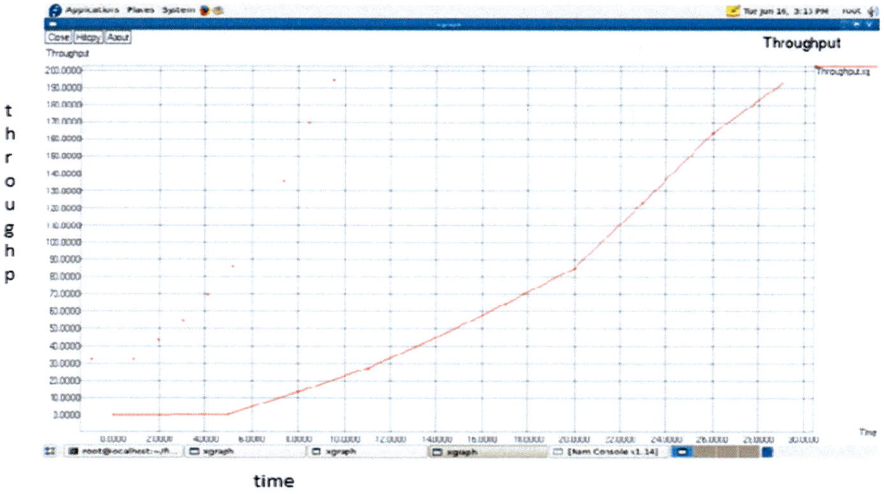

Fig. 3. Time vs throughput of DERDWSN

In Fig. 6 it can be seen Energy consumption in proposed protocol is comparatively less than AODV and DSR. Table 2 shows comparison of Existing and Proposed Protocol and percentage improvement. In above mentioned QoS parameters, we can observe that our research concern about the energy efficient parameters through which we can proposed our work for data transfer over wireless network. If it will be implement in real practical application then we will get better result through our proposed protocol.

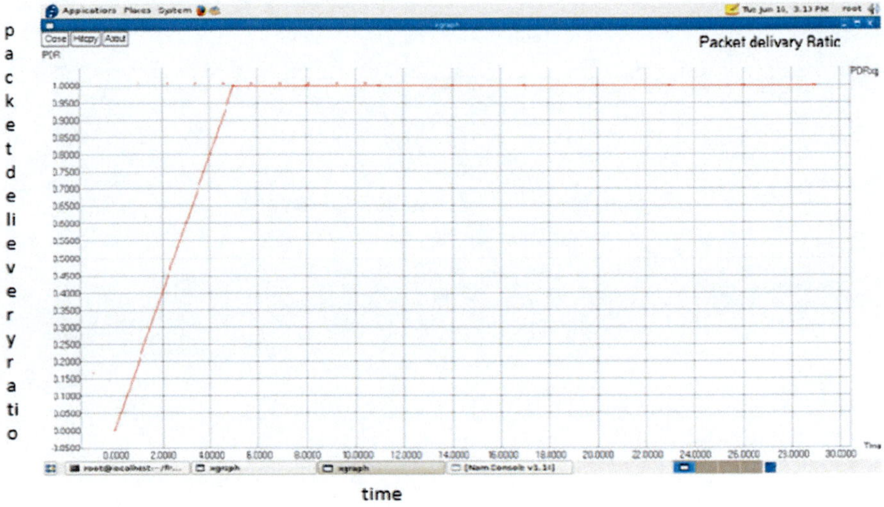

Fig. 4. Time vs packet delivery ratio of DERDWSN

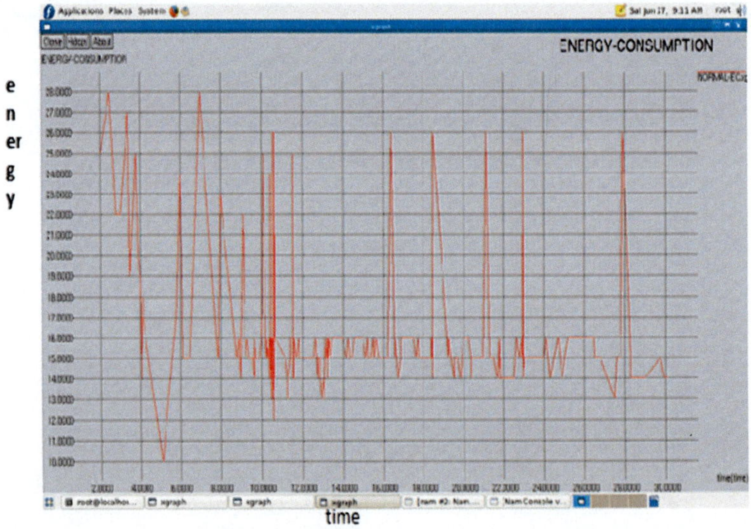

Fig. 5. Time vs energy consumption of DERDWSN

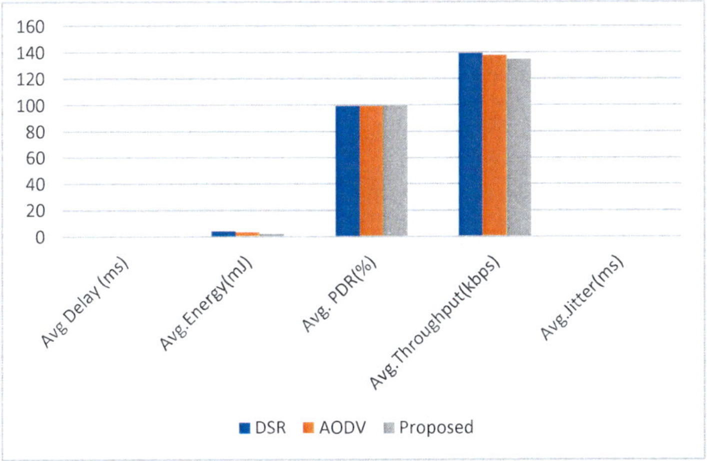

Fig. 6. Comparison of DSR, AODV and proposed protocol

Table 2. Comparison of existing and proposed protocol

Parameter	DSR	AODV	Proposed	% Improvement
Avg. Delay (ms)	0.65	0.556	0.412	26%
Avg. Energy (mJ)	4.30	3.26	2.198	29%
Avg. PDR (%)	99.3	99.5	99.6	0%
Avg. Throughput (kbps)	139.3	137.8	134.9	– 0.50%
Avg. Jitter (ms)	0.0069	0.0062	0.0058	7%

5 "Conclusion"

We concluded from the simulation results derived from the given model that DERDWSN, PDR and performance are lower at energy consumption, which is an important element in the performance of the wireless sensor network. This reduces the usage of protocols based on energy consumption clusters. From the present simulation model and the above results are obtained. It is reflected that proposed method for designing the protocol for data transfer over wireless network is superior to the previous network simulation. It improves the performance the following parameter e.g. throughput, PDR, Jitter, and most important is energy efficient based network in which data transfer through cluster head is very easy and safety.

References

1. Abbasi AA, Younis M (2007) A survey on clustering algorithms for wireless sensor networks. Comput Commun 30:2826–2841

2. Sohrabi K et al (2000) Protocols for self-organization of a wireless sensor network. IEEE Pers Commun 7(5):16–27
3. Min R, et al (2001) Low power wireless sensor networks. In: Proceedings of international conference on VLSI design, Bangalore, India, January 2001
4. Akyildiz IF, Su W, Sankarasubramaniam Y, Cayirci E (2002) Wireless sensor networks: a survey. Comput Netw 38(4):393–422
5. Heinzelman W, Chandraksan A, Balakrishnan H (2002) An application specific protocol architecture for wireless micro sensor networks. IEEE Trans Wirel Commun 1:660–670
6. Muruga Nathan SD, Ma DCF, Bhasin RI, Fapojuwo AO (2005) A centralized energy-efficient routing protocol for wireless sensor networks. IEEE Radio Commun Mag 43:8–13
7. Tang F, You H, Guo S (2010) A chain-cluster based routing algorithm for wireless sensor networks. J Intell Manuf 23:1305–1313
8. Samia A, Shreen K (2011) Chain-chain based routing protocol. IJCSI Int J Comput Sci 8(3):105
9. Bian X, Liu X, Cho H (2008) Study on a cluster-chain routing protocol in wireless sensor networks. In: The 3rd international conference on communications and networking, China
10. Xiangning F, Yulin S (2007) Improvement on LEACH protocol of wireless sensor network. In: Proceedings of the international conference on sensor technologies and applications, USA

Check for updates

Real Time ModBus Telemetry for Internet of Things

T. Shiyaz(✉) and T. Sudha

ECE Department, NSS College of Engineering, APJ Abdul Kalam Tehnological University, Palakkad, Kerala, India
shiyaz61@gmail.com, sudhat@nssce.ac.in

Abstract. The Internet of Things is a system in which sensors and actuators can send and receive data across a network and connect with one another [1]. Modbus is a standard communication protocol widely used in industrial automation. Modbus Protocol is a messaging structure used in multiple client server applications to monitor and program devices and to communicate between intelligent devices and sensors. In this paper, an inexpensive pilot setup using Modbus TCP communication for the Internet of Things applications in real time device telemetry using embedded systems is modelled. This proposed pilot setup can able to collect device data through modbus protocol and able to provides the connectivity extension of Modbus devices to IoT applications by performing local data processing. In order to extend the existing modbus devices to IoT have to rely on external data converter's or industrial gateway [2]. In market there are ModBus IoT gateways are available however we can able to achieve the same goal using the proposed pilot setup. The proposed pilot setup developed using inexpensive embedded boards can able to perform modbus real time telemetry through its local data processing it can able to collect data from their existing devices through their modbus protocol and able to extend to IoT for real time telemetry.

Keywords: Internet of Things · Modbus protocol · Embedded systems · M2M communication · Device telemetry

1 Introduction

Machine to machine, M2M is exchanging data or communicating between two machines without human interactions. M2M and Internet of Things (IoT) are almost synonyms, the exception is IoT generally refers to wireless communications and M2M refers communication between two machines can be often of wired or wireless. The Internet of Things is a system in which sensors, and actuators are prepared to send and receive data via a network as well as connect with one another [3]. In order for machine to communicate to each other a communication protocol is required. Most industrial systems uses proprietary communication protocols and it cannot provide interoperability between two machines of

R. Misra et al. (Eds.): ICIoTCT 2021, LNNS 340, pp. 129–138, 2022.
https://doi.org/10.1007/978-3-030-94507-7_13

different vendors. However, the MODBUS TCP is an open standard and simple communication protocol used for many industrial automation. Modbus is based on a polling mechanism and follows the synchronous request-response pattern [4].

This paper describes the real time device telemetry system developed using Embedded Systems for Internet of Things applications. The goal is to design an inexpensive pilot setup for M2M communication using Modbus for real time device telemetry to extend the connectivity of Modbus world to IoT world by performing local data processing. The test was carried out in network interconnecting MODBUS Server, client and slave device and MODBUS TCP was employed for synchronous polling communications [5]. The implementation of MODBUS server and client is based on Linux operating system. Modbus exchange starts from a client, when the client sends a request to the server to execute a command. Modbus Server will provide response to the client request by providing the requested data.

The content of this paper is organized as follows: Sect. 2 describes about quick overview of the ModBus Protocol. Section 3 focus on the proposed system implementation. Section 4 focuses on test setup results and Sect. 5 concludes the paper and presents the planned future ideas.

2 Modbus Protocol

2.1 Modbus Overview

Modbus is a communication protocol developed by the Modicon Corporation. It's widely used in industrial automation. Using the Modbus protocol, manufacturers' control devices or measuring instruments can be integrated for industry monitoring [6].

Table 1. ModBus data types

Object type	Access	Size
Discrete input	Read-Only	1-bit
Coils	Read-Write	1-bit
Input registers	Read-Only	16-bit
Holding registers	Read-Write	16-bit

The Table 1 represents the modbus data model. The data blocks consists of 4 groups known as Discrete Input, Coils, Input Registers and Holding Registers. Modbus registers will be located in the device application memory. The function codes defines the action to be performed by the Modbus server [7].

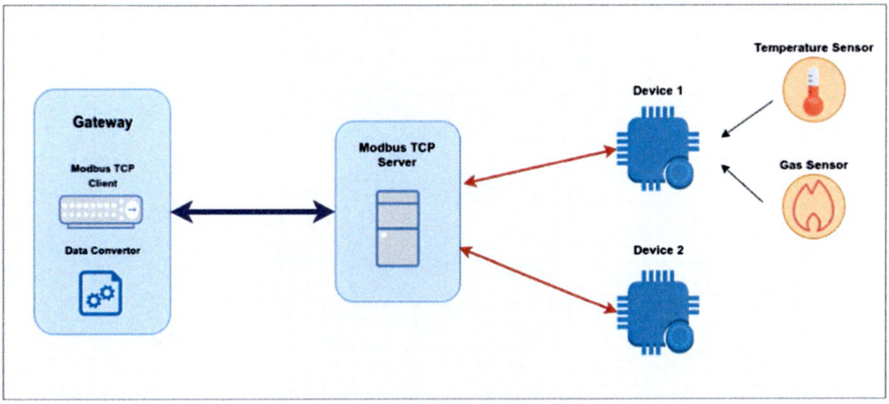

Fig. 1. Test setup

3 System Description

The Fig. 1 shows the pilot setup block diagram. The MODBUS exchange of information between server and client. MODBUS Request will be initiated by the Client to initiate a transaction, MODBUS Response message is sent by the Server and MODBUS Confirmation is the Response Message received on the Client side.

3.1 Pilot Setup

The Fig. 2 and Fig. 3 shows the test setup for the real time modbus telemetry with embedded devices and sensors connected. It consists of two MODBUS Slave devices on that Slave I is a development embedded board which consists of Temperature & Humidity Sensor and Gas Sensor connected and Slave II is a simulated dummy device it is a virtual device. The MODBUS TCP Server, which basically manages the slaves which are connected to the server and it will perform actions initiated by the MODBUS TCP Client. The exchanges of information between a Modbus client and server will starts by client sends a request to the server to execute a command and after the server receives the request, it executes or retrieves the required data from its memory. Then the server will respond to client by providing the requested data by the server. The slaves will send sensors telemetry values to MODBUS TCP Server through Serial communication at frequent time interval. In this case the sensors values are temperature, humidity and Gas value. The Modbus TCP Server will collect those values from serial communication and maintain updates ModBus Slaves Memory registers. Modbus TCP Client also frequently queries for the latest Temperature & Humidity value from the Modbus TCP Server.

Fig. 2. Pilot setup

Fig. 3. Pilot setup with temperature, humidity and gas sensor

The components used for this pilot setup as follows:

- Raspberry Pi Model B development embedded board which was used for modbus server.
- Nvidia Jetson Nano development kit board which was used for modbus client.
- NodeMCU ESP8266 dev kit which was used for slave devices and Temperature & Humidity and Gas sensors which was connected to slave devices through UART connections.

The test procedure followed for this pilot setup as follows:

- Connected Temperature and Humidity DHT-11 Sensor & Gas Sensor to embedded board serially through UART connection.
- Established a connection between Modbus server and client through Ethernet network interface and connection between Modbus server and slave devices through UART connections.
- The slave devices which holds the sensors which will send sensor telemetry values to the Modbus Server through serial communication.
- Modbus Server updates its slaves memory registers as per latest telemetry received from slave devices.
- Modbus server service can be started so that server can be able to respond to clients request.
- Modbus client service can be start at a time interval loop so that client can query for the latest device telemetry values for the time interval.

4 Test Results

4.1 Serial Communication

Fig. 4. UART serial communication

The Fig. 4 shows the serial console of a MODBUS slave device which has temperature and humidity sensor connected. This slave will send temperature and humidity value to the master through serial communication UART communication protocol.

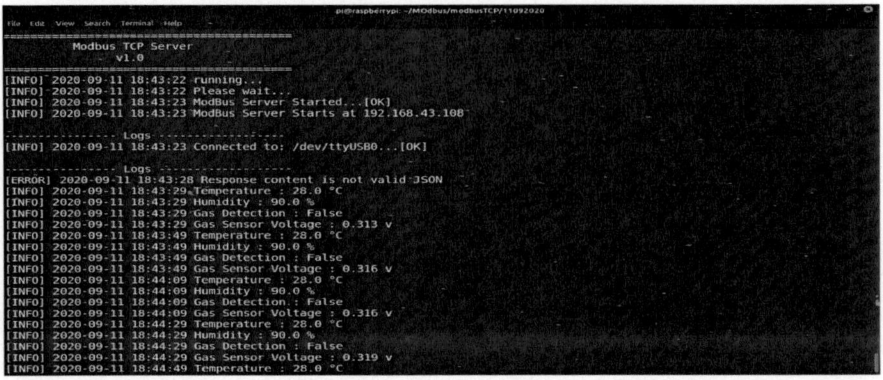

Fig. 5. Modbus TCP server

4.2 ModBus TCP Server

The Fig. 5 shows the serial console of a MODBUS server. This server is running on a embedded device. The Modbus server which hold the resources and will collect real time temperature, humidity and gas readings from the sensors through serial communication and updates its memory registers according to the real time values. The memory registers always hold the latest telemetry value fetched from the sensors.

4.3 ModBus TCP Client

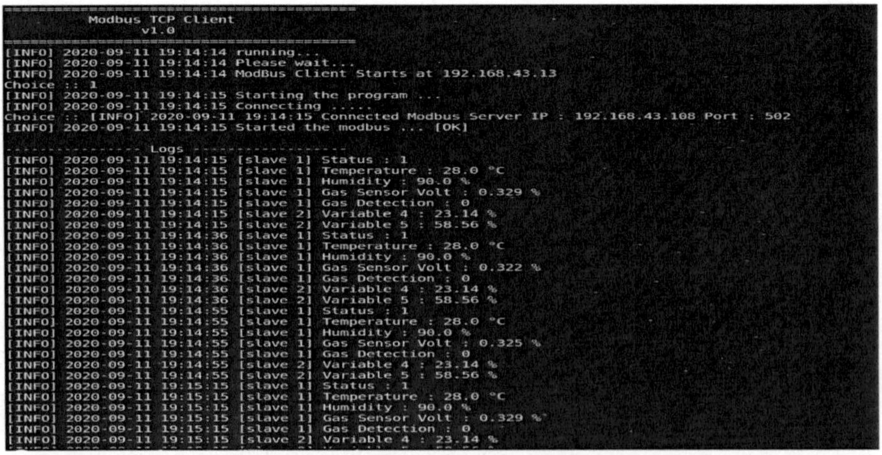

Fig. 6. Modbus TCP client

The Fig. 6 shows the serial console of a MODBUS client. The client is also running on a embedded device. The exchanges of information between a Modbus client and server will starts by client sends a request to the server to execute a command and after the server receives the request, it executes or retrieves the required data from its memory. Then the server will respond to client by providing the requested data by the server. Modbus Client will query temperature and humidity value over 10 s interval from the MODBUS TCP server in order to get the latest telemetry values.

4.4 Network Analysis

Wireshark is the worlds popular open source network protocol analyzer and licensed under GNU General Public License (GPL) Version 2. This tool can able to capture the live network packets and provides details of the packets to debug the protocol implementations. The test procedure followed to capture the live MODBUS packets as follows.

- Wireshark Packet Analyzer was setup up in the Modbus Client to capture live network packets between server and client with choosing Ethernet as network interface to capture packets from.
- After the network interface selection, the packets can be captured by selection start in capture section. This action will start Wireshark to capture all incoming and outgoing packets from Ethernet interface.
- Wireshark also comes with a handy feature known as capture filters by applying filters we can limit the packet capturing to capture only the ModBus protocols traffic.
- Wireshark provides a range of network statistics which can be used to plot graphs and plots from captured packets.
- After stopping the live network capture the captured live modbus packets can be dumbed as hex format and can be able to saved in a single file in the .cap file format and which can be use for the future references.

The Fig. 7 shows captured modbus packets using Wireshark network packet analyzer.

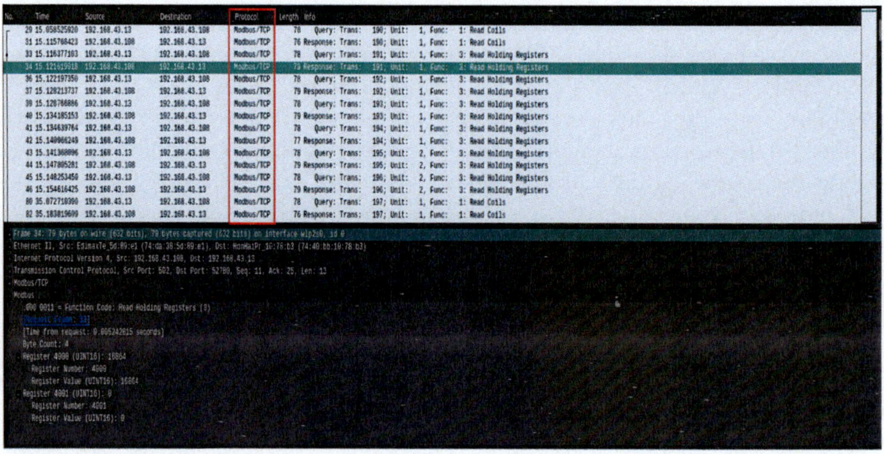

Fig. 7. Analyzing packets through packet analyzer

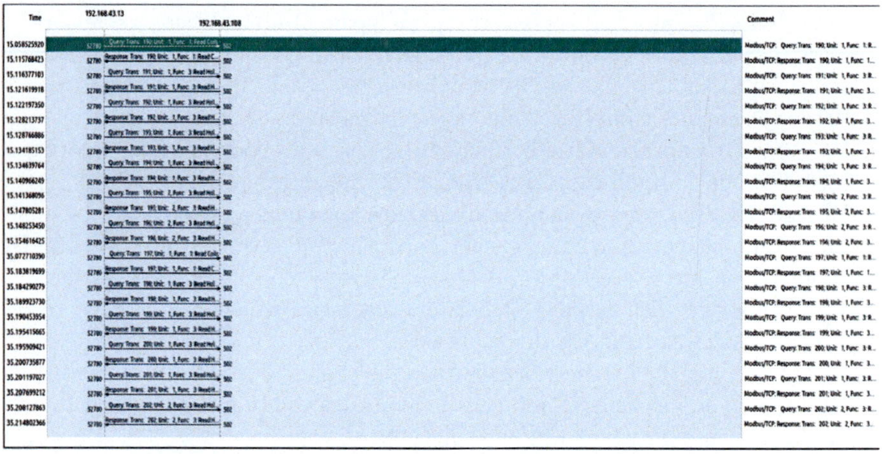

Fig. 8. Network flow graph

Wireshark has a handy feature that is display filter through filter modbus packets can be selected for network analysis. The Fig. 8 shows network flow graph plotted from wireshark network packet analyzer. The flow graph shows the connection between the hosts. It can also display packet time, direction, ports and comments for each captured packet. In this case Modbus server is having IP address of *192.168.43.108* and Modbus client having the IP address of *192.168.43.13*.

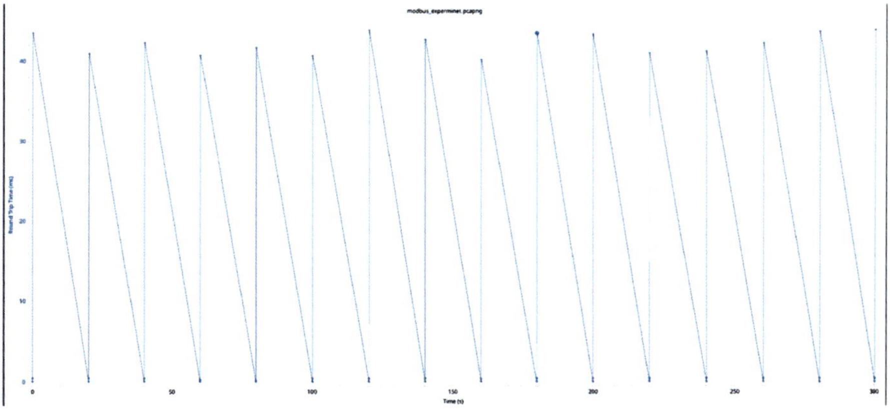

Fig. 9. Round trip time

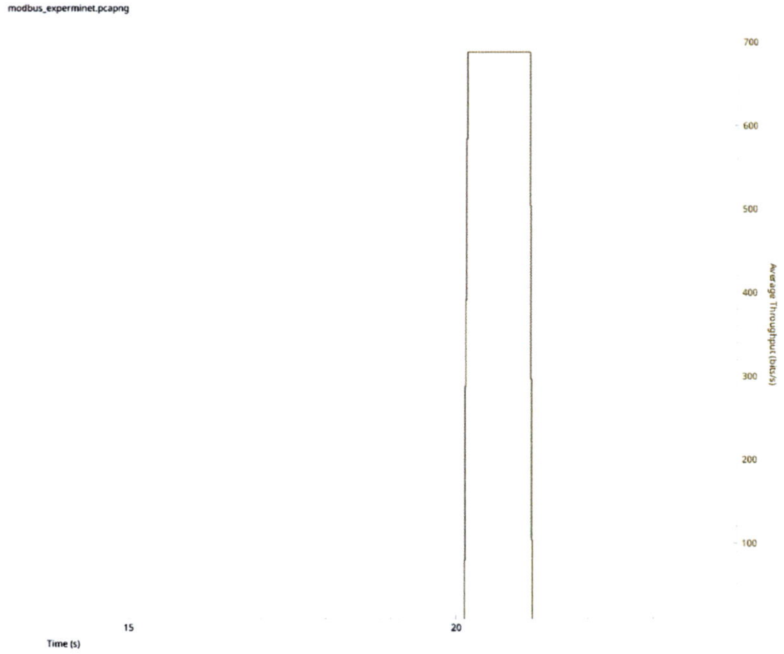

Fig. 10. Average network throughput

The Figs. 9 and 10 shows Round trip time (RTT) & Average network throughput plotted from Wireshark network packet analyzer statistics tool. RTT is based on the acknowledgement timestamp corresponding to a particular segment.

5 Conclusion

This test aims at the possibility of Modbus network for Internet of Things applications with real time telemetry. The test was carried out in network interconnecting MODBUS Server, client and slave device and MODBUS TCP follows the synchronous polling communications. This test can provide the feasibility of M2M communication using MODBUS TCP for real time telemetry in Internet of Things applications. The MODBUS master can query the sensors data from the server at frequent intervals. The implementation of MODBUS server and client is based on Linux operating system. Modbus TCP connection can be achieved from a low cost development boards that supports TCP/IP and also it is simple to implement for any device that supports TCP/IP sockets. In theory Modbus TCP/IP carries data about 60% efficiency and the theoretical throughput is equal to 360000 registers per second [7]. The essential goal of IoT is to utilize the power of communication or internet connectivity and computation to pre-existing or in-use real-world objects by using an existing network [1]. An average of 45.2 ms of RTT and average network throughput of 700 bits per second could be achieved from this test setup. However, network infrastructure, network traffic and physical distance between server and client are all important factors that can affect the RTT. Further work will include development of a web application, an gateway software platform to control the system with real time communication which makes it more efficient and intelligent.

Acknowledgment. This work is supported by Kerala Startup Mission, Govt. of Kerala through the funding for establishing of IoT lab to Innovation and Entrepreneurship Development Centre. (IEDC), NSS College of Engineering, Palakkad, Kerala.

References

1. Ray AK, Bagwari A (2020) IoT based smart home: security aspects and security architecture. In: 2020 IEEE 9th international conference on communication systems and network technologies (CSNT), pp 218–222. IEEE
2. Kuang Y (2014) Communication between PLC and Arduino based on Modbus protocol. In: 2014 fourth international conference on instrumentation and measurement, computer, communication and control, pp 370–373. IEEE
3. Nugur A, Pipattanasomporn M, Kuzlu M, Rahman S (2018) Design and development of an IoT gateway for smart building applications. IEEE Internet Things J. **6**(5) (2019)
4. Shu F, Lu H, Ding Y (2019) Novel Modbus adaptation method for IoT gateway. In: 2019 IEEE 3rd information technology, networking, electronic and automation control conference (ITNEC), pp 632–637. IEEE
5. Sun C, Guo K, Xu Z, Ma J, Hu D (2019) Design and development of Modbus/MQTT gateway for industrial IoT cloud applications using Raspberry Pi. In: 2019 Chinese automation congress (CAC), pp 2267–2271. IEEE
6. Modbus-IDA (2006) Modbus messaging on TCP/IP implementation guide V1.0b, North Grafton, Massachusetts. www.modbus.org/specs.php
7. Modbus-IDA (2015) Modbus application protocol specification V1.1b, North Grafton, Massachusetts. www.modbus.org/specs.php

The Link Between Emotional Machine Learning and Affective Computing: A Review

Utkarsh Singh(✉) and Neha Sharma

CSE Department, Indian Institute of Information Technology Una, Una, India
neha@iiitu.ac.in

Abstract. This paper explores the relation between Emotional Machine Learning, a research area that involves making Neural Networks learn using various learning methods of the mammalian brain that involve emotions; and Affective Computing, a field that is concerned with recognizing, interpreting, processing and simulating human emotions. Both these areas find a common point, that is simulating human emotions. While we are far from truly making a machine that could replicate emotions in Affective Computing, developments such as Emotional Backpropagation Algorithm, BELBIC, etc. could play a pivotal role in directing our Affective Computing applications from building affective interactive chatbots to affective learning systems that perform better than standard learning systems. Both these research areas can benefit from individual developments of their counterparts in advancing their own field. In this paper, we review the progress in Emotional Neural Networks and BELBIC models, and see how they relate to Affective Computing.

Keywords: Emotional backpropagation algorithm · Learning theory · Emotional neural networks · Affective computing

1 Introduction

Before moving onto the discussion of how emotions have been added into the field of Machine Learning and Intelligent Systems, we need to first reason the necessity or benefits of the pursuit. Due to the blurred line of applications of this idea, Hollnagel [5] has considered the term "affective computing" a "brainless phrase" because a field should refer to a specific use of computing rather than a type of computing. There are various arguments that assert that enhanced empathy in machines could lead to better interaction with humans when it comes to chatbots and voice interaction softwares. Picard claims that the Turing Test can be passed by a machine if it is made capable of perceiving and expressing emotions [2].

There have been huge strides in perceiving emotions based on text, visuals and audio signals. However, it is the expression of emotions by a machine that sparks a good debate. Khashman has stated and proves that we are still not sure about what truly is the essence of emotions in the human mind [1]. The void of knowledge here limits us to provide that essence to a system too. This means that whatever progress is made without the

R. Misra et al. (Eds.): ICIoTCT 2021, LNNS 340, pp. 139–147, 2022.
https://doi.org/10.1007/978-3-030-94507-7_14

knowledge of the essence of emotions, the system is purely mimicking the way a human would react.

However, Stromfelt [6] has outlined numerous ongoing researches that are getting closer to understanding the true origination and process of emotions, and how they can be mathematically simulated. Fellous [9] noted that emotions in the brain originate from the amygdala, that stores the stimuli-emotional response and provides emotional coloration to declarative memories. However, the process of emotion generation is not entirely restricted to this section of the brain. Yet, numerous models like BELBIC and BEL were still developed which simulated certain chunks of processes of the brain [13].

As long as we do not find the complete working of the brain with respect to emotions, however complex and far-fetched it may seem, it seems futile to model the brain's sub-systems to generate emotions by taking away incomplete knowledge of unitary structures of the brain for a process that is clearly involving multiple structures.

2 Discussion

2.1 Emotional Backpropagation Learning Algorithm

Khashman proposed this algorithm in 2008, as a way to model certain human emotions in a neural network. Contrary to the previous debate of whether adding emotions could make the interaction better, the entire purpose of adding emotions here is different. Lewin [10] notes that even in humans, the role of emotions is not just to display empathy, but also to make rational decisions, and people who suffer from damages in the emotional centers of their brain actually are unable to make proper decisions. This had been originally noted by Picard [2] too.

Khashman has not dived deep into the inner workings of the brain and borrowed a chunk of the brain's emotion elicitation process. Instead, he has borrowed a chunk of our emotional learning behavior. The crux of the emotional backpropagation (EmBP) algorithm is that the network is given an anxiety level and a confidence level. As the network learns more and adapts to the data, it gains more confidence and its anxiety is lowered. This behavior is inspired from how a human learns. This modified neural network was tested against a general neural network for the task of facial recognition and it was found to be performing better.

When tested against a basic neural network, the EmBP based neural network had a performance increase of 3%, that is, it correctly recognized a greater number of images than the novel neural network. Interestingly though, the training time, run time and number of iterations were all greater in case of the EmBP based NN than the BP-based NN.

Analysing the algorithm through the perspective of the original philosophy that states that anxiety is higher at the start of learning, if we were to say that the algorithm is mimicking the human way of learning and the current way is better, the human element of this algorithm is actually slowing down the training process, but making the decisions much better.

Plausible factors that could influence the success of the algorithm could be the order in which the data is fed to the network, the change in training and testing images and the variance in coefficients of anxiety and confidence (Fig. 1).

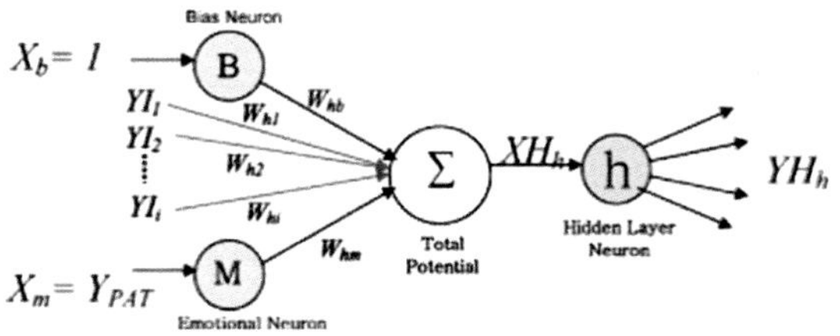

Fig. 1. Schematic diagram of a neuron with the emotional neuron [1]

Each hidden-layer neuron's output is defined as:

$$YH_h = \left(\frac{1}{1 + \exp(-XH_h)}\right) \tag{1}$$

Here, X_h is defined as:

$$XH_h = TP_{hc} + TP_{hb} + TP_{hm} \tag{2}$$

Where TP_{hc} is the conventional total potential from the previous layer neurons and their associated weights, TP_{hb} is the potential obtained from the bias neuron and its associated weights and finally TP_{hm}, is the potential obtained from the emotional neuron and the weights associated with it.

TP_{hc}, TP_{hb}, TP_{hm} are defined as:

$$TP_{hc} = \sum\nolimits_{i=1}^{r} W_{hi}.YI_i \tag{3}$$

$$TP_{hm} = W_{hm}.X_m \tag{4}$$

Here, X_m is defined as:

$$X_m = Y_{PAT} \tag{5}$$

where Y_{PAT} is taken as the input image's P (x, y) global input pattern average value:

$$Y_{PAT} = \sum\nolimits_{x=1,y=1}^{x_{max},y\_max} \frac{P(x,y)}{x_{max}.y\_max} \tag{6}$$

where x_{max}, y_{max} are the total number of pixels in the x and y axes of image P (x, y), respectively.

Two parameters, namely the anxiety coefficient and the confidence coefficient, that vary between 0 and 1, have been proposed, with a few specifications. Similar to human behavior, newer patterns shall cause higher anxiety. Thus initially, anxiety coefficient is 1 (highest). The anxiety is also dependent on the difference between the desired (target) output and the actual output of neural network. Anxiety is proportional to this difference (error). Confidence coefficient increases with the decrease in anxiety coefficient.

$$\mu = Y_{AvPAT} + E \tag{7}$$

Here, Y_{AvPAT} is the average of all the presented patterns to the neural network in each iteration:

$$Y_{AvPAT} = \sum_{p=1}^{N_p} Y_{PAT}/N \tag{8}$$

where p is a pattern index from first to the last pattern. The error feedback is defined as:

$$E = \sum_{j=1}^{N_j} \frac{\left(T_j - YJ_j\right)^2}{N_p.N_j} \tag{9}$$

The new weights are updated taking into account the change in conventional weights and the change in the emotional weights (Table 1).

$$W_{jh(new)} = W_{jh(old)} + \Delta W_{c_{jh}} + \Delta W_{m_{jh}} \tag{10}$$

The change in conventional weights is:

$$\Delta W_{c_{jh}} = \eta.\Delta_j.YH_h + \alpha.\left[\delta W_{jh(old)}\right] \tag{11}$$

And the change in emotional weights is:

$$\Delta W_{m_{jh}} = \mu.\Delta_j.Y_{PAT} + k.\left[\delta W_{jh(old)}\right] \tag{12}$$

Table 1. EMBP and BP based neural networks

Training images	Testing images	Accuracy of EmBP-based NN	Accuracy of BP-based NN
200	200	90%	90%
160	240	87%	84%
120	280	84%	81%

Use of ENN in Facial Recognition

Facial Recognition was used to test the efficiency of the EmBP Algorithm. A face image database was provided to a standard backpropagation Neural Network and an emotional backpropagation Neural Network, with the same parameters wherever possible, and the results were compared.

It was clear that while the number of iterations to converge was higher than the conventional BP algorithm (EmBP: 6680, BP: 4813), the model trained by EmBP performed better than the one trained by conventional BP. There might be scope for a few more unconventional testing techniques, given the fact that this algorithm is changing the way Neural Networks learn. Certain ratios of Training-Testing databases might give a fairly better performance than other ratios and there could be a change in the convergence iterations depending on the random initialization of parameters and order of the facial image data, since anxiety and confidence would vary depending on the next average pattern value of the image.

Table 2. Neural networks for credit evaluation

Neural network models	Performed iterations	Hidden layer nodes	Learning coefficient	Momentum rate	Learning scheme	Dataset accuracy	Runtime (x 10^{-5} s)	Error
NN-1	25000	10	0.0075	0.9	LS1	77.18%	4.11	0.0161
NN-2	13841	9	0.00935	0.67	LS1	79.23%	4.11	0.007
NN-3	25000	10	0.0085	0.87	LS2	77.97%	9.28	0.0093
NN-4	6292	9	0.009	0.88	LS2	79.42%	9.28	0.007
NN-5	25000	10	0.0077	0.89	LS3	79%	15.67	0.0224
NN-6	16592	9	0.0068	0.90	LS3	80.33%	15.67	0.007
EmNN-1	11834	10	0.0075	0.9	LS1	81.03%	3.85	0.007
EmNN-2	25000	9	0.00935	0.67	LS1	78.72%	3.85	0.0172
EmNN-3	17615	10	0.0085	0.87	LS2	80%	8.99	0.007
EmNN-4	25000	9	0.009	0.88	LS2	79.71%	8.99	0.0162
EmNN-5	8407	10	0.0077	0.89	LS3	77.67%	5.33	0.007
EmNN-6	25000	9	0.0068	0.90	LS3	80.67%	5.33	0.0199

2.2 Testing Emotional Neural Networks for Credit Risk Evaluation

In a separate paper, Emotional and Conventional models of Neural Networks were used to classify credit risk evaluations [4]. 6 different neural networks, each of conventional and emotional models, were used to predict if a credit application will be approved or declined. 14 numerical attributes from an Australian credit approval dataset totalling 690 cases were used to train the 12 neural networks.

The 6 neural networks for each model were arbitrated on the momentum rate, the count of hidden layer nodes, and the learning coefficient.

Although the difference was not that significant, experimental results did suggest that the standard backpropagation algorithms did not perform as well as the emotional models in decision making rate and precision. It must be noted that while training, the average number of iterations to converge while training was higher for emotional neural networks than for conventional ones. Although this highly depends on the random initialization of the weights, the higher number of iterations and greater training time can be attributed to the fact that there are additional parameterized computations quite different than the conventional way. Table 2 lists the results of twelve different neural networks trained under three different ratios of training to testing data, also known as learning schemes. (Training: Testing- 300:390, 345:345, 390:300). The emotional neural models (EmNN-1, EmNN-3 and EmNN-5) and conventional models (NN-2, NN-4 and NN-6) converged to the required error value of 0.007 within 25,000 iterations, which were the maximum allowed iterations. Of these six neural models, the model with the quickest runtime is EmNN-1, with a run time of 3.85×10^{-5} s and accuracy of 81.03% on the validation dataset.

2.3 Prototype-Incorporated Emotional Neural Network

In cognitive science, in a mode of categorization, few members of a notional class are kept more central than others. This graded categorization is also known as Prototype theory [8]. The adaptive-learning theory explains that learning can be accomplished using numerous instances of the target to be studied to regulate internal parameters of the model; this is in contradiction to the prototype-learning theory. Khashman and Oyedotun have proposed a new Neural Network model which aims to combine both adaptive- and prototype-learning theories in a neural network. Since adaptive learning theory is traditionally the basis of Neural Networks, this can also be seen as prototype learning being incorporated into conventional neural networks. It is also built as an addition to the EmBP algorithm proposed by Khashman earlier. The reason for this is that the authors have tried to simulate the emotional response from the emotion neuron on to the correlation and prototype neuron weights too.

This paper proposes 2 additional neurons in addition to Khashman's Emotional Neuron in EmBP [1]: A prototype neuron denoted by P, that provides the normalized prior prototype class label of the given input data to the network's output and hidden layers, and a correlation neuron denoted by C, that feeds both the output and hidden layers the correlation coefficient of the input data.

Prototype Neuron (P)
There are two approaches to having the prototype neuron:

1. One Prototype Per Class Approach: The input data has various classes. From each set of patterns in a class, an example is selected at random to form the prototypes. A distance metric is now used to get the closest prototype to the training data (attributes).

$$d = \sqrt{\sum_{u=1}^{n} \left(x_u - x_u^P\right)} \tag{13}$$

   x_u – the input attribute from the given training data where u is the index.
   x_u^P – the input attribute from the given prototype where u is the index.
   n - the dimensionality of the given input patterns and prototypes.
2. Multi prototypes Per Class Approach: From each class, a few prototypes are randomly selected from the given training data.

Correlation Neuron (C)
Pearson's Correlation Coefficient R is used to define C = R2. R is defined as:

$$R = \frac{\sum_{u=1}^{n}\left(\left(x_u - \overline{x_u}\right)\left(x_u^P - \overline{x_u^P}\right)\right)}{\sqrt{\sum_{u=1}^{n}\left(x_u - \overline{x_u}\right)^2 \sum_{u=1}^{n}\left(x_u^P - \overline{x_u^P}\right)2}} \tag{14}$$

Here, x_u, x_u^P and n are the same as defined before.

Activation Computations

The weights of the prototype neuron in the hidden-to-output layer are updated by: δw_{kp} is the change in weight for the hidden-to-output layer prototype neuron.

$$\omega_{kp}(i+1) = \omega_{kp}(t) + \mu\Delta_k P + k\left[\delta\omega_{kp}(i)\right] \tag{15}$$

Similarly, the hidden-to-output layer correlation neuron weights are updated using

$$\omega_{kc}(i+1) = \omega_{kc}(t) + \mu\Delta_k C + k[\delta\omega_{kc}(i)] \tag{16}$$

Application of the PI-EmNN to Hand Gesture Recognition and Face Recognition

The new proposed Neural Network model was tested along with the conventional Neural Network and Emotional Backpropagation-based Neural Network. The various PI-EmNNs used are PI-EmNN1, PI-EmNN3 and PI-EmNN5 where the number represents the number of prototypes per class.

The following table lists the final results of the various networks for the tasks of Static Hand Gesture Recognition and Face Recognition (Tables 3 and 4).

Table 3. Final results for face recognition

Network	PI-EmNN1	PI-EmNN3	PI-EmNN5
Training time (s)	643.8	657.2	668.5
Mean squared error (MSE)	0.0038	0.0025	0.0017
Anxiety coefficient	0.0054	0.0050	0.0049
Confidence coefficient	0.6187	0.6221	0.6253

Table 4. Final results for static hand gesture recognition

	TT(s)	MSE	Anxiety coeff.	Confidence coefficient
PI-EmNN1	564.7	0.0035	0.0321	0.2798
PI-EmNN3	586.49	0.0030	0.0097	0.4946
PI-EmNN5	584.16	0.0077	0.0108	0.3427
EmNN	544.31	0.0057	0.0123	0.04201
BPNN	504.47	0.0058	-	-

2.4 Beyond Emotional Neural Networks

Stromfelt had summarized, on the basis of several conferences that the emotion is fundamentally produced in the amygdala [6], instead of the limbic subsystem of the brain in humans, and specifically noted in [11, 12].

However, to encompass the research that has been going on to connect emotions in the brain, we shall look at BELBIC, a model of computation which mimics the limbic subsystem of the mammalian or human brain for control engineering applications.

BELBIC (Brain Emotional Learning Based Intelligent Controller):

With reference to the schematic structure of BELBIC given below, the formula for the learning rule of amygdala is:

$$\Delta G_a = k_1 . \max(0, EC - A) \quad (17)$$

The orbitofrontal cortex's learning rule is defined as:

$$\Delta G_0 = k_2 .(MO - EC) \quad (18)$$

Here, G_0 - the change in orbitofrontal connection,

k_2 - the learning step in orbitofrontal cortex

MO = A – O. MO signifies the output of the whole model, and O is the output of orbitofrontal cortex (Fig. 2).

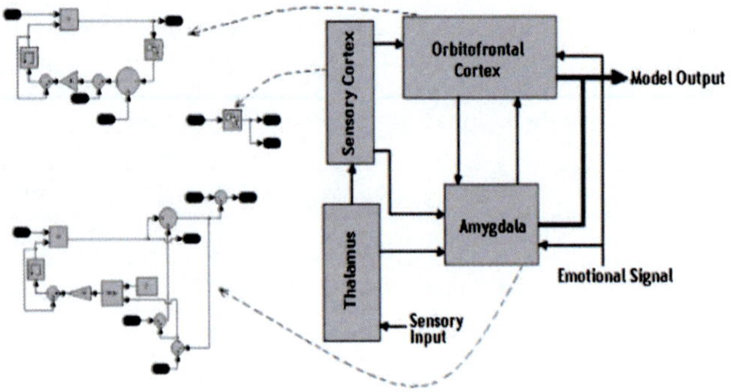

Fig. 2. Schematic structure of the BELBIC model

By obtaining the sensory input, S, the model can compute the amygdala's and orbitofrontal cortex's internal signals using the equations below.

$$A = G_a.S \quad (19)$$

$$O = G_o.S \quad (20)$$

The amygdala is unable to unlearn an emotional response. It is the duty of the orbitofrontal cortex to inhibit any inappropriate response.

The functions that are used in sensory input blocks and emotional cue are given in below equations:

$$C = W1.e + W2.CE \quad (21)$$

$$SI = W3.PO + W4.PO \quad (22)$$

Here EC, CE, SI and PO stand for emotional cues, control effort, sensory input and plant output. W1, W2, W3 and W4 are the gains that must be tuned for building a suitable controller.

3 Conclusions

In this paper, we reviewed many algorithms and learning models that have tried to mimic human emotions and human learning and scratched a certain surface of Affective Computing. While it may be argued that the tests and models of Affective Computing cannot be related with the EmNNs and BELBIC models above, it is important to note that simply demonstrating that a computer has emotions by a Turing Test is not, and should not be enough, as outlined by many essays. Rather, emotions in machines should be used to produce quality outputs, improving the computations that predominantly govern our world. Currently, simulating emotions has proved an increase in performance, which signals that actually building models replicated from the mammalian emotional brain structure could provide something extraordinary.

References

1. Khashman A (2008) A modified backpropagation learning algorithm with added emotional coefficients. Trans Neural Netw 19(11):1896–1909
2. Picard RW (1997) Affective computing, vol 252. MIT Press, Cambridge
3. Martinez-Miranda J, Aldea A (2004) Emotions in human and artificial intelligence, Elsevier, Amsterdam
4. Khashman A (2011) Credit risk evaluation using neural networks: emotional versus conventional models. Elsevier, Amsterdam
5. Hollnagel E (2003) Is affective computing an oxymoron? Elsevier, Amsterdam
6. Stromfelt H, Zhang Y, Schuller BW (2017) Emotion-augmented machine learning: overview of an emerging domain. In: 2017 seventh international conference on affective computing and intelligent interaction (ACII). IEEE
7. Oyedotun K (2017) Prototype-incorporated emotional neural network. IEEE Trans 29:3560–3572
8. Ehsan Lotfi M-R, Akbarzadeh T (2014) Practical emotional neural networks. Neural Netw 59:61–72
9. Fellous J-M, Armony JL, Ledoux JE (2002) Emotional circuits and computational neuroscience. Neuroscience 454(7200):1–8
10. Lewin D (2001) Why is that computer laughing? Histories and future. IEEE Intell Syst 16:79–81
11. LeDoux JE (2000) Emotion circuits in the brain. Annu Rev Neurosci 23(1):155–184
12. Fellous J-M, Armony JL, Ledoux JE (2002) Emotional circuits and computational neuroscience. Neuroscience 454(7200):1–8
13. Lucas C, Shahmirzadi D, Sheikholeslami N (2004) Introducing Belbic: brain emotional learning based intelligent controller. Intell Autom Soft Comput 10(1):11–21

Check for updates

Arduino Based Temperature, Mask Wearing and Social Distance Detection for COVID-19

Jash Shah(✉), Heth Gala, Kevin Pattni, and Pratik Kanani

Department of Computer Engineering, Dwarkadas J. Sanghvi College of Engineering, Mumbai, India
shahjash271@gmail.com, hethgala@gmail.com,
kevinbpattani@gmail.com, pratikkanani123@gmail.com

Abstract. With the world experiencing a paradigm shift in its working culture and people working from home, while it's been beneficial to many industries helping them with cost-cutting, some have suffered irrationally at the hands of COVID-19, and for them, employees returning to site has become the need of the hour, with lockdown rules easing out and new strains of viruses being discovered, utmost precaution needs to be taken. Hence there is a need for a comprehensive system that can ensure proper protocols being followed as well as manage the office workloads for each and every employee which is what we'll be discussing about in detail, the various libraries, tools, frameworks, machine learning algorithms, sensors used to realize a Zero Human Contact cost-efficient IoT-Machine learning-enabled setup capable of managing the whole office and ensure a safe working environment for its staff.

Keywords: IoT · Attendance tracker · Social distancing detection · Arduino · YOLOv3 · COVID19 · RESNET-30

1 Introduction

The whole world is struggling to combat COVID-19. Albeit, it can possibly take several more years for the virus to be annihilated completely. Time of paramount importance since offline work needs to resume as quickly as possible for affected sectors. In large scale multinational companies, with hundreds of employees per branch, it can be quite difficult to keep a check on the attendance and maintain social distance at the same time.

With safety precautions being enforced by the need of the Hour (Notably Social Distancing, Wearing Mask) we plan to create a system that can not only efficiently handle office workloads and optimize productivity within office Spaces but also enforce that all covid-19 precautions are being followed. The system as such would be divided into 2 phases:

a) The First phase will be responsible for Detecting the Person has worn a mask and has temperature in the permissible range to allow entry and mark Attendance.
b) The second phase will be responsible for Detecting if social distancing violations are being followed or not.

R. Misra et al. (Eds.): ICIoTCT 2021, LNNS 340, pp. 148–161, 2022.
https://doi.org/10.1007/978-3-030-94507-7_15

The only common aspect in both of these phases from the point of view of technology is the application of machine learning. Here we are rigorously using one of the specialized branches of Deep Learning – Computer Vision. It helps us in object detection and face recognition, two of the most important facets of our system. The system will be realized by making use of IoT.

2 Literature Review

Several works exist in automated attendance systems [14, 15] and social distancing violations [16]. Face recognition technique using FACECUBE components is proposed [14]. The FACECUBE components transfer the intense work of detection and recognition to the server-side computer.

Face recognition technique using OpenCV and Light Toolkit is proposed [15]. The face information is stored in an xml file and eigenvalue and eigenvector is calculated for that image. Finally, the face image is matched with existing face images in xml files.

Both the above research works implement face recognition which is not very ideal in COVID scenarios considering wearing masks is mandatory and our system ensures wearing it. Our system first scans if there is a face, checks if there is a mask put on and then only if there is a mask on, the QR Code is scanned. This scanning process isn't very demanding in terms of computation, which enables the existing client hardware sufficient for the purpose.

Additionally, these systems [17] which do implement these systems have been implemented on R-Pi, the downside to this approach is that R-Pi is very costly and difficult to configure the first time thus we chose to make use of Arduino in order to overcome these difficulties.

Conventionally face mask recognition models have been created using the primitive fit() method which is time and cost inefficient and hence has been tackled well in our use case wherein we have gone on to use fit_one_cycle() of Fast Ai which provides better accuracy alongside time and cost efficiency [18]. Several works have been done where a social distancing detector has been implemented. Social Distancing and Safety Violation Alert using OpenCV and Caffe framework is proposed [16]. This approach however is computationally costly due to multithreading and prone to deadlock situation at the same time, thus we have avoided going down that path and have rather taken an approach that's computationally optimized and fast, utilizing yoloV3 to create bounding boxes around the subject and accordingly calculate the distance between the subjects - frame by frame and avoiding multithreading and consequently avoiding any chances of deadlock whatsoever. The Subject enters at the Gate, where the system checks, if the person has worn a mask or not and whether the temperature is in the permissible range. If both the criteria is not met, then entry is denied, else the user is prompted to generate the QR code on the Application to Mark for attendance by placing it in front of the scanner, if the QR code is Valid attendance is Marked and Entry is Granted. Once in the Office, social distancing violations are logged and the employees are warned

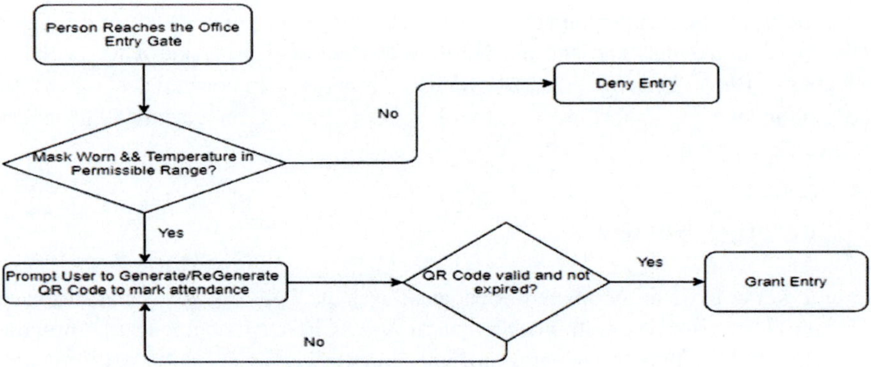

Fig. 1. System diagram

3 Components Used

We have made Use of Arduino Uno board since it is cheap and very easily available and easy to interface with other sensors used. Given the Ease of interfacing and cost, we have made use of MLX90614 as a temperature sensor, Radio Frequency (RF) modules for communication between Arduino Uno's.

3.1 Arduino Uno

Arduino is an open-source platform that allows to create programmable electronic devices that can be interfaced with various other sensors. Arduino Uno [2] is an ATMEGA 328 Chip based Microcontroller. It is the most widely used microcontroller of the Arduino Family. Its Features Include:

- Input Voltage required is between 6-20 V
- There are 14 Digital I/O pins
- There are 6 Analog Pins
- Tx,Rx pins are used for serial communication
- It has an USB interface

In order to program the Board, a dedicated IDE is required that transfers the code from the system (in which the IDE is) to the Arduino board. It can be programmed in either C or C++ or can also be interfaced with other languages making use of additional extensions and libraries. Power is supplied to the board by making use of a battery or an AC/DC Adaptor. Its an open source, a lot of support and libraries are available to interface Arduino with multiple sensor languages, etc. In our case we will be making use of 2 Arduinos, one that ensures protocols are followed and marks attendance and controls the gate. The 2 Arduino's will communicate with each other via RFID [4] (Fig. 2).

Fig. 2. Elements of arduino uno board

3.2 Temperature Sensor (MLX90614)

MLX90614 is an IR temperature sensor used for non-contact measurements, that's extended out towards the user using an extension. It is the same component that's used in thermal Guns and has a very high accuracy and precision.

Its Features are:

- Operating Voltage is between 3–5 v
- Object Temperature Range: −70 °C–382.2 °C
- Ambient Temperature Range: −40 °C to 125 °C
- Accuracy: 0.02 °C

It is very easy to interface with Arduino and is already factory calibrated hence it is as easy as plug and play, requiring no extra components. The sensor as whole is divided into 2 parts (Fig. 3 and Fig. 4).

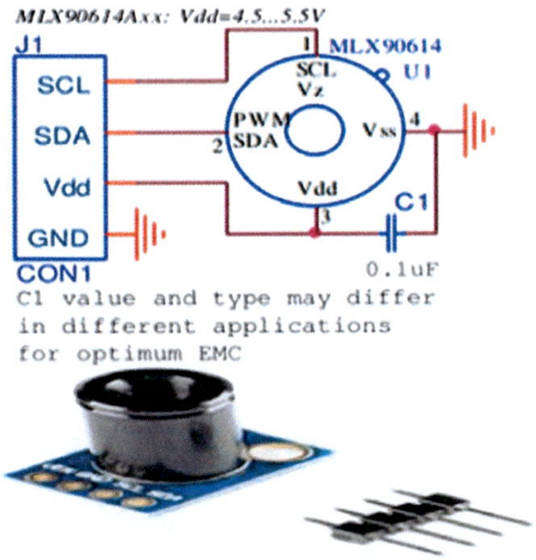

Fig. 3. Temperature sensor

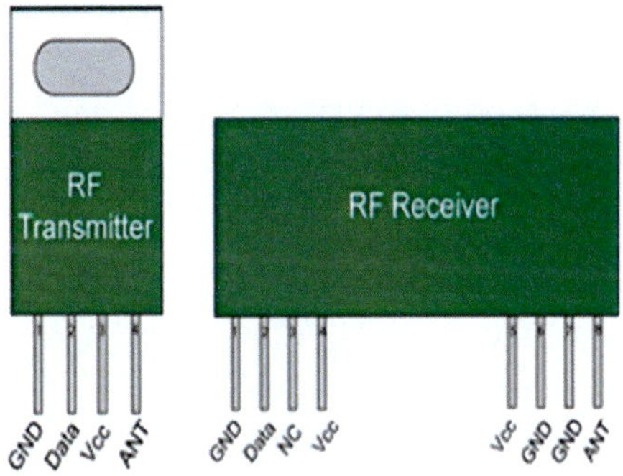

Fig. 4. RF receiver and transmitter

- **Sensing Part** - Infrared Thermopile Detector called MLX81101 which senses the temperature
- **Processing Part** - Single Conditioning ASSP called MLX90302 which converts the signal from the sensor to digital value and communicates using I2C protocol [3].

Radio Frequency (RF) [4].

a) **Transmitter** - A radio frequency (RF) signal refers to a wireless electromagnetic signal used as a form of communication. An RF module is a small electronic device used to transmit and/or receive radio signals between two devices. It consists of 2 parts: Transmitter and Receiver.
b) **Receiver**: -It is used as a component to receive data in radio signals

Features: -

- Receiver Frequency is 433 MHz
- It has an operating voltage of 5 V
- Transmitter: -It is used for transmitting data in radio signals, through its antennae connected at Pin4Features: -
- It operates in Frequency Range - 433.92 MHz.
- It requires a supply voltage of: 3 V–6 V.

4 Methodology

System Overview

The Whole solution setup can be split into 2 parts:
Outside Office (Pre-Entry)
Inside Office
Outside Office Setup - The Outside office setup can be thought to be split into 2 parts

- Detection Module
- Attendance Taker

Detection Module: Here the Arduino is Interfaced with Webcam, making use of the PySerial Library (It is a python package that is used to interface Arduino with python Programming Language) The Reason for choosing python to interface with Arduino was because the detection and recognition model was developed with python, it was easier to run the model in python programming language only rather than in C/C++. Additionally, OpenCV (it is an open-source computer vision and machine learning library) Computer Vision [5].

As the name suggests, this section of learning aims to help computers "see" and understand the contents of digital images such as photographs or videos. The computer vision problems seem trivial or petty to humans sometimes, as this is a common notion because we humans excel at computer vision tasks. For example, identifying simple objects like cat, dog and even tiger and lion who lie under the cat family. This seems simple but when it comes to training the computer for these tasks, we need a lot of data to explain the nuances of the images and the objects we clearly aim to recognize correctly. The primary aim of computer vision is to extract useful information from the images or photographs been provided but despite the efforts from the last four decades we are still far from being able to build a general purpose "sewing machine".

In our case we require computer vision to help us by recognizing the person's face and then by recognizing the mask on that facial structure if present. Here we use computer vision to break down our images to the three primary layers like red, green and blue layers which are effectively the matrices of these three colors building up to make the complete image. Then on these three sets of color matrices the algorithm is run to effectively yield a final set of matrices that can be matched with a matrix that we yield after sheer training on the data set (the set of images which look alike the input image, in our case the images of people wearing masks) (Fig. 5).

A small example of the data set that is being provided for our module. Here in Fig. 1, we see Fig. 6 images labeled as mask or no-mask. So, the dataset not only contains the true data but also the odds so that machine also learns from the odds and well recognizes such odds when presented as a new input. This shall help our computer learn from these images and recognize the defaulters at the doorstep of the office and restrict their entrance into the premises. Machine learning and training models to do their work accurately is a tedious task. To ease out some of those tasks we have used Fast Ai which is a deep learning centered library. Hence it is used to spin up the webcam and capture Frames which are fed to the python script that makes use of the OpenCV function that is used to reference the Webcam Cap = cv2.VideoCapture(0). This will start the Webcam ret,img = cap.read() Here the captured frame is assigned to the image variable.

Fig. 5. Data samples of dataset used

Fig. 6. Example of the detection and the recognition model that detects and predicts if mask has been worn or not

Step I: Now we can pass this image variable to the function that makes use of the model to detect if the subject has worn a mask or not.

The module of mask detection has been made possible using Fast Ai [6] library. Let us learn more about **Fast Ai:** As the name suggests it makes you Ai development models work faster than usual. This is a library exclusively built on another famous library for Deep Learning: PyTorch. It has gained a lot of popularity in recent years for its advancements, code conciseness, effectiveness and primarily developer friendly. Thus, Fast Ai has gained a lot of popularity too as it makes PyTorch even more effective and faster. It is a library which is used to make Computer Vision models and Neural Networks in Machine Learning. Thus, we have chosen Fast Ai for our mask detection module making the best use of the library PyTorch and making our detections faster than usual. Fast is the new normal in the 21st era of technology boom thus we aim to make all the segments of this project as fast as possible so that one has to wait for minimalistic time which shall be faster than what the world provides today.

Hence, we have chosen Fast Ai to ease out the work of computer vision at a high accuracy rate and concise code structure. Therefore, it is developer friendly too. Let us discuss about how the whole procedure works in the Office Spaces Project: At the first stage where the employee undergoes the checks like temperature check and mask detection, if the temperature is within the normal range and the mask is worn in a feasible manner then only the Arduino allows the employee to enter the office opening the gates at the entrance itself.

The mask detection module uses the Computer Vision module provided by Fast Ai which is mobilenetv2 by Google. Let us iterate through stepwise according to the code written to make mask detection possible:

1.1) For every supervised learning algorithm a dataset is quite necessary for it to work well. So firstly, we load in the dataset which consists of around 16,200 images in.png format for our model to train on. Dataset provided is derived from Kaggle: "ahmet furkan demir/mask-datasets-v1".

1.2) The dataset thus loaded needs to be cleaned or made in a format to be fed into the algorithm at hand.

- At first, we split the total dataset into the "Train" set and the "Cross-Validation" set. The train set is the one which we feed in the algorithm whilst training it, i.e. the algorithm shall learn from the fraction of pictures which we got from Kaggle while the Validation set is the one which is used to test our algorithm. It's just like teaching a child a topic in school and then to evaluate his knowledge, the child undergoes a test. The same is the case here, the Validation set acts as a test for us to evaluate the level to which the model is ready to recognize the given entity (in our case the "mask").
- After the split we go through the transform step which is specific to Fast Ai library whereby data is converted to a desired size and a data bunch of such images is created. data bunch is a jargon used in the Fast Ai world. In this process we shall also normalize the dataset. Now normalization is a process to convert the columns that are the entities in our dataset to a common scale, avoiding any changes in the ranges of value or losing any information provided.

1.3) After the dataset is ready to be fed in the algorithm, now we need the algorithm itself. Here we are using the Fast Ai method cnn_learner() to load in the algorithm in our module. Loading the already made algorithm with a fixed metric of accuracy and error rate is the intermediate process in our module. As compared to other models MobileNet v2 is more memory efficient and faster to train as compared to Resnet50, Resnet18, VGG19, which are more complicated architecture and their trained models take up a memory space of about 100 mb as compared to MobileNet's 20 mb which makes it easier to deploy. Hence we shall be using mobilenetV2, i.e., the version 2 of the newly formed Google's MobileNet.

Let us dive deeper to know what is MobileNet v2 [7] and how it works:

MobileNet was introduced into the world of science by Google researchers who intended to bring Machine Learning to the Mobile world without compromising accuracy and at the same time minimizing the number of parameters and mathematical operations required. This is purely because mobile devices cannot take that much load to do vigorous calculations and analysis and this led to the birth of MobileNet.

This architecture makes use of the concept of depth-wise disparate convolutions containing depth-wise as well as point-wise convolutions sequentially. MobileNetV2 was introduced to adjust the MobileNet world to the small but vivid range of mobile devices.

It is well known as an inverted residual structure where the output as well as input of the Convoluted Neural Network structure are residual blocks of thin bottleneck layers which is quite opposite to the ones used in the traditional method which used expanded representations in the input while V2 uses lightweight depth wise convolution to filter features in the intermediary layers. Inverted Residuals [8]. Earlier the residual blocks were connected from the beginning of CNN to the end by a skip connection (Fig. 7 and Fig. 8)

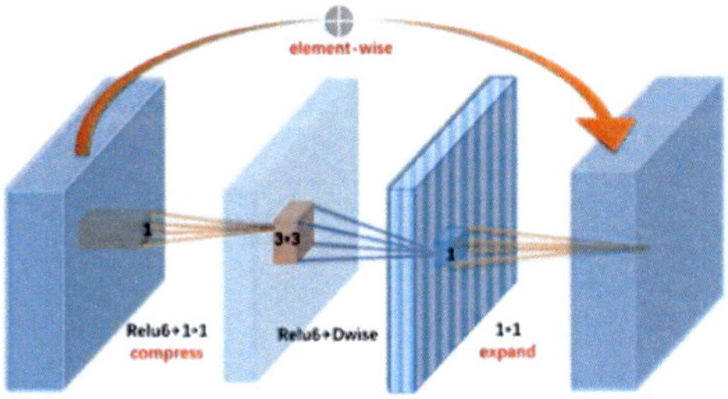

Fig. 7. Residual block

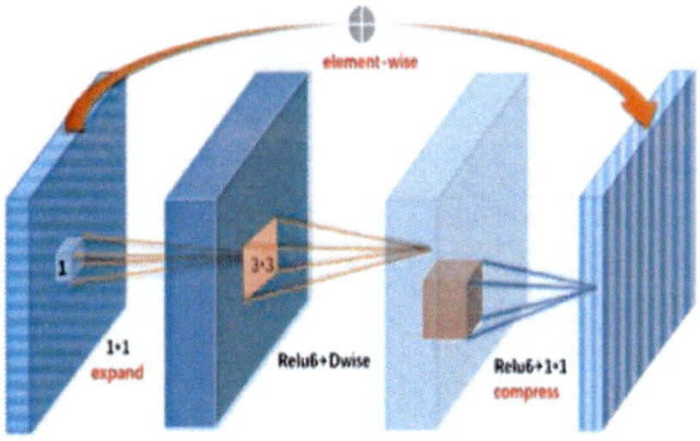

Fig. 8. Inverted residual block

The motive behind adding these states to the network is to obtain data with respect to earlier activations that remained unchanged. It is a wide-narrow-wide approach. The high number of channels passed as input are compressed using a 1×1 convolutional layer. In this way 3×3 convolution has far less parameters. In the end the last layer is again increased using 1×1 convolution.

In MobileNetV2 a narrow-wide-narrow approach is adopted. The first being to widen it using a 1×1 convolution layer because the number of parameters is drastically reduced in a 3×3 depth wise convolutional layer. Afterwards 1×1 convolutional is again used to squeeze to match the original input at the end. As the skip connections that exist between minuscule gaps at the end of modules that are in nature antagonistic to traditional approaches are thus known as inverted residual blocks.

Linear Bottlenecks: In CNN we often use the ReLU activation function which rejects negative values. This loss of values is overcome by improving the capacity of the network which happens when the number of channels is increased. Now linear bottleneck in this case helps tackle the condition of inverted residual blocks where the residual block's last convolution gives a linear output before adding it into the initial activations.

ReLU6: ReLU6 serves to add Batch Normalization to every convolutional layer in the architecture. Now let's see what Batch Normalization is. Normalization is a process whereby the input layer is adjusted by scaling the activations. We have some features in the dataset provided which may range from 0–1 and some which may range from 1–1000 and some which even wider range. So, to give equal priority to all the entities and bring them to a common scale for analysis we do normalization. It brings all the entities to a single scale and helps in effective calculations. So now what is basically Batch Normalization, is a process where we normalize not only the input but also the intermediary neural network layers/hidden layers to increase the performance and time efficiency which is 10 times or more increased training speed of the model. The difference between ReLU and ReLU6 is that it limits the value of activations to a maximum of 6. That indicates that the activation remains linear as long as it's within the range of 0 to 6. The figure indicates ReLU6 clearly.

1.4) Now we have a better understanding of what MobileNetV2 is and a brief understanding of the layers that work within it, we can proceed to the calculations of the cnn_learner. This learner takes in the data as its first parameter which is primarily our data bunch as discussed earlier (Fig. 9 and Fig. 10).

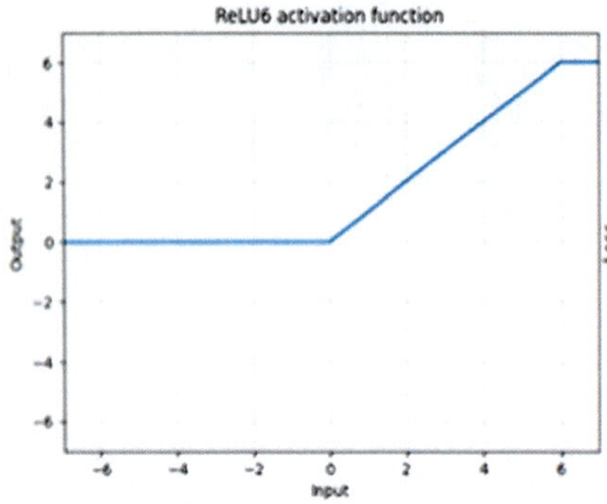

Fig. 9. ReLU activation function.

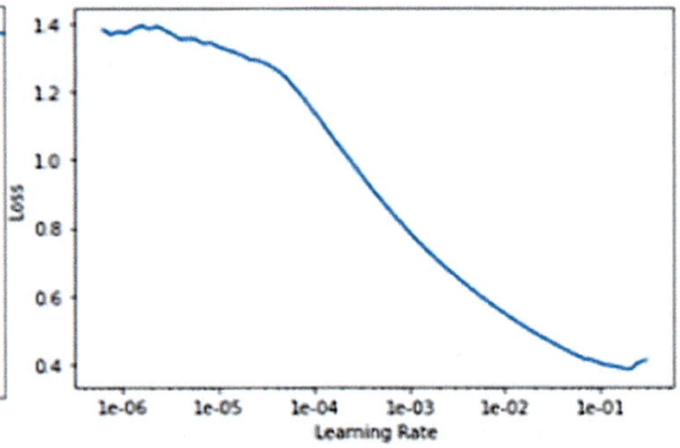

Fig. 10. Finding the optimal learning rate.

The second param is the mobileNetV2 model which gives us the precalculated activations on the module. The third and one of the most important parameters is the metric. A metric is one on the basis of which we decide the performance or the training of the model that we shall train. In our case we input the accuracy and error rate as the metric parameter. The find() method used in the code trains the model and returns back the train and validation loss with our two metrics the accuracy and the error rate. We also get back the graph showing Loss versus the Learning Rate. The ideal graph for a model is the one with lowest loss with an optimal learning rate. The graph in our case looks somewhat like this: We can regulate our parameters and for one extra cycle to train our model to increase our accuracy levels using fit_one_cycle(), it uses large, cyclical learning rates to train models significantly quicker and with higher accuracy. On getting the desired accuracy for our model, i.e., 99.71% we can stop regulating and training any further, to save the weights thus calculated for future validations.

1.5) Retrieving the confusion matrix is one important step in our procedure to get a pictorial representation of our accuracy. In our case the matrix looked somewhat like this: Here as we see the actual true masks detected were 149 and false detections were just 1. Whereas true no mask detections were 200 and false were 0. This itself speaks of how accurate our model is (Fig. 11).

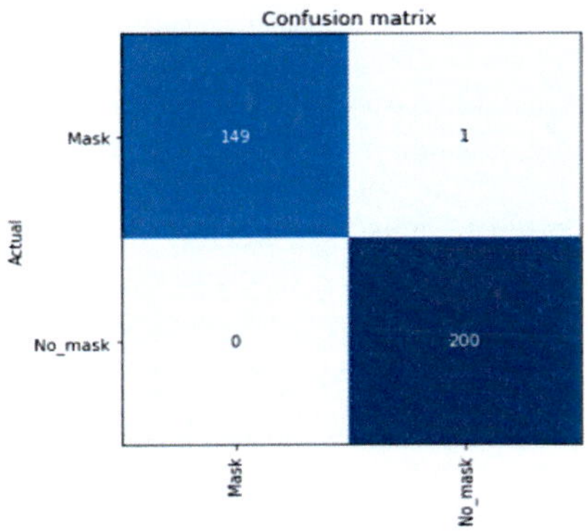

Fig. 11. Confusion matrix

1.6) We finally start testing dummy images with and without masks to see how our model performs on the data which it has never encountered. The prediction can be done by inputting the image into the predict() function provided by Fast Ai.

This marks the ending of the total work flow. We retrieve results indicating whether the person has worn a mask or not in the form of tensor values which indicate the confidence of the algorithm in indicating which is the correct output. The confidence of the algorithm in detecting masks or no-masks helps us clearly indicate the same and conditionally open the doors at the office gate for the employee.

On the basis of this, function returns a Boolean value 1 or 0.

0: Denoting the subject hasn't worn a mask.

1: Denoting the subject as a worn mask.

On the basis of this Boolean value the python script then sends across 1 or 0 byte encoded to Arduino.

Step II: Arduino receives this data using the serial Port, following which Arduino makes use of Temperature Sensor (MLX90614) in order to measure temperature

If temperature is in the permissible range (as per covid-19 norms) then make use of the transmitter mounted on the breadboard to send across the signal to the other Arduino which has the receiver interfaced with it.

In order to leverage full capability of using the transmitter given the fact we have more than one sensor interfaced with Arduino, we make use of < RH_ASK.h > header, which is essentially a RadioHead library and create its rf_driver object that's initialized in the Void Setup () loop of Arduino.

Step III: The User is then prompted to scan his QR code (generated from his mobile app) The Second Arduino again is interfaced with Python to make use of OpenCV library to leverage webcam that serves as a QR code scanner.

Step IV: Once the frame is extracted it is passed to a function that decodes the QR Code, in order to decode QR code (the information stored in it) by making use of PYZBAR library [9].

Step V: Once the information is extracted from the QR code a subsequent API call is made to the backend server to verify the user based on the decoded data information.

Step VI: The server queries across its database to find the targeted user and queries against the current time and decoded time stamp to ensure the QR code has not expired.

Step VII: Once the user is verified his/her attendance is marked in the database and the Intime is logged, the server sends back a Boolean response, which is passed onto the Arduino board and then the user is accordingly prompted to enter if everything's proper (Fig. 12 and Fig. 13).

Fig. 12. Arduino 1 - Transmitter setup, that detects mask, checks temperature and sends signal to Second Arduino

Fig. 13. Arduino 2 - Receiver Setup that is interfaced with QR scanner to mark attendance

Inside Office Setup –
The inside office module consists of two sections:

1) Social Distance Tracker:
YOLOv3: You only look once (YOLO) is a state-of-the-art, real-time object detection system. On a Pascal Titan X, which is one of the best graphics card money can buy, YOLOv3 [10] has a MAP (mean average precision) of 57.9%. It has a comparable performance to SSD and RetinaNet which was considered as one the best object detectors, but is now replaced by YOLOv3 which is said to be 3.8 times faster in image recognition than all RetinaNet subnetworks (ResNet, FPN, FCN). It is an improvement on YOLOv2 which has 30 layers in its Darknet-19 architecture, whereas this has Darknet-53 which

has 106 total layers in its network making it a bit slower in detection than others but increasing its accuracy. We can see how the YOLOv3 has performed on the COCO dataset with respect to other detectors in the graph given below [11].

One of the main reasons we have selected YOLOv3 is because of IoU. IoU (Intersection over Union) is one of the most important aspects we must take into consideration before choosing the object detection model. It means how much the object's boxes will overlap, the more the overlap the higher the IoU [12] and vice versa.

Since, in social distance we are mostly going to face low IoU, we have chosen YOLOv3 as it is the fastest at lower IoU than any other object detectors.

YOLOv3 Darknet: The predecessor of YOLOv3 Darknet, YOLOv2 initially used a 30 layered network in which 11 layers were concatenated later for the sole purpose of Object Detection. Even after this modification YOLOv2 still struggled with it (minute object detection). It precipitated the down sampling of input leading to loss of minute features. To solve this, the newer version of YOLOv2 tried to capture the detailed features by merging the feature maps of the previous layer. Many state-of-the-art algorithms were already using residual blocks, skip connections and up-sampling which was still lacking in the upgraded version of YOLOv2. To suffice the lacking features Joseph Redmon and Ali Farhadi came up with Yolov3. YOLOv3 is a modified mutant of Darknet, which initially used a 53 layered network which is trained on ImageNet. Specifically for the task of detection additional 53 layers were stacked upon the original Darknet model, giving it a robust and efficient 106 layered convolutional architecture for YOLOv3. It was found in a study that YOLOv3 darknet-53 proved to be 2 times faster than ResNet-152. In our Office Spaces project YOLOv3 works in a way such that, whenever our camera detects a person YOLOv3 creates a bounding box around it (the box mentioned earlier to describe IoU is a bounding box) which defines the space which the object (here the person) covers. Once the boxes are created their centroid is calculated and then it becomes the effective point from which all the distances are calculated. Now, let's assume there are two people in the vision of the camera. So, a bounding box is created around each person and their respective centroids are created. Now using those centroids as two points we created, we calculate the distance between them and if the distance is less than what it should be, then it is a social distancing violation (Fig. 14 and Fig. 15).

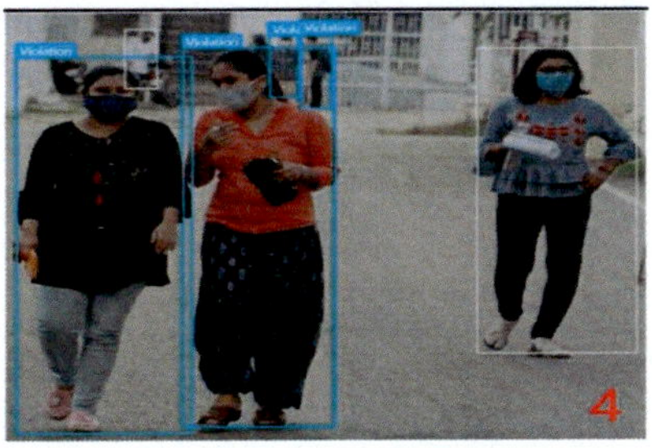

Fig. 14. Social distancing, violation detected

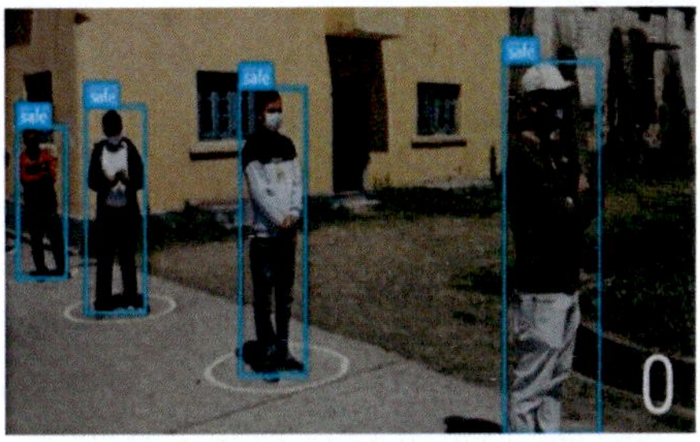

Fig. 15. Social distance detection

2) Facial Recognition for Mask Violators

Face Recognition (face recognition and Siamese networks): To learn about face recognition using Siamese Networks, we need to have knowledge about One-Shot Learning Techniques [13]. These are basically techniques that allow us to develop good machine learning models using just a few training examples. Neural Networks are very good at solving almost every problem provided that there is enough data to train the model properly, increase its accuracy and improve its predictions. But it's plausible that while encountering some problem statements, we might face the issue of data shortage and to get we resort to solutions like synthetic data or even data generation which can be quite expensive and might lead to a simpler dataset which does not help our model evolve. This is where the Siamese Networks come into the picture. They are the most efficient and simple One-Shot Learning Techniques.

5 Conclusion

Hence, we have realized a system that is Cost-effective, portable and makes sure that people follow precautions, leveraging the power of IoT (making use of Arduino) in order to detect if the subject has worn a mask and enforce social distancing. The software has been successfully implemented and has been done so in such a modular manner that each and every aspect of the project can be implemented by itself in an isolated environment without having any interdependency between these modules whatsoever, at the same time capable enough to be integrated all together. The system has been extended keeping a long-term view in mind, wherein the scope of the project is not just limited till the time covid-19 is to exist but can also be utilized as an ERP system that can help improve employee productivity - right from attendance tracking to handling internal office activity, at a single place, rather than them having to make use of 10 different applications to achieve the same end result effectively.

References

1. InDepth View of IOT Architecture. https://www.hiotron.com/iot-architecture-layers/
2. Kanani P, Mamta P (2020). Real-time location tracker for critical health patients using Arduino, GPS Neo6m and GSM Sim800Lin health care. In: 2020 4th international conference on intelligent computing and control systems (ICICCS), pp 242–249. https://doi.org/10.1109/ICICCS48265.2020.9121128
3. MLX90614 Non-Contact IR Temperature Sensor. https://components101.com/sensors/melexis-mlx90614-contact-less-ir-temperature-sensor
4. RF Module – Transmitter & Receiver. https://www.elprocus.com/rf-module-transmitter-receiver/
5. Culjak D, Abram T, Dzapo PH, Cifrek M (2012) A brief introduction to OpenCV. In: 2012 proceedings of the 35th international convention MIPRO, Opatija, pp 1725-1730
6. fastai: A Layered API for Deep Learning. https://arxiv.org/abs/2002.04688
7. MobileNets: Efficient convolutional neural networks for mobile vision applications. https://arxiv.org/abs/1704.04861
8. MobileNetV2. Inverted Residuals and Linear Bottlenecks. https://arxiv.org/abs/1801.04381
9. Pyzbar. https://pypi.org/project/pyzbar/

10. Lee Y, Lee C, Lee H, Kim J (2019) Fast detection of objects using a YOLOv3 network for a vending machine. In: 2019 IEEE international conference on artificial intelligence circuits and systems (AICAS), Hsinchu, pp 132–136. https://doi.org/10.1109/AICAS.2019.8771517
11. Yolo v3-Object Detection. https://syncedreview.com/2018/03/27/the-yolov3-object-detection-network-is-fast/#:~:text=At%20320%%20%2020x%20320%2C%20YOLOv3,on%20a%20Pascal%20Titan%20X
12. IoU a detection evaluation metric. https://towardsdatascience.com/iou-a-better-detection-evaluation-metric-45a511185be1
13. Melekhov I, Kannala J, Rahtu E (2016) Siamese network features for image matching. In: 201623rd international conference on pattern recognition (ICPR), Cancun, 2016, pp 378–383. https://doi.org/10.1109/ICPR.2016.7899663
14. Godswill O, Osas O, Anderson O, Oseikhuemen I, Etse O (2018) Automated student attendance management system using face recognition. Int J Educ Res Inf Sci 5(4):31–37
15. Kar N, Deb Barma MK, Saha A, Rudra Pal D (2012). Study of Implementing automated attendance system using face recognition technique. Int J Comput Commun Eng 1:100–103. https://doi.org/10.7763/IJCCE.2012.V1.28
16. Ahamad H, Zaini N, Latip MFA (2020) Person detection for social distancing and safety violation alert based on segmented ROI. In: 2020 10th IEEE international conference on control system, computing and engineering (ICCSCE), 2020, pp 113–118. https://doi.org/10.1109/ICCSCE50387.2020.9204934
17. Ruhitha V, Prudhvi Raj VN, Geetha G (2019) Implementation of IoT based attendance management system on raspberry PI. In: 2019 international conference on intelligent sustainable systems (ICISS), Palladam, India, pp 584–587. https://doi.org/10.1109/ISS1.2019.8908092
18. Explaining adv of fit_one_cycle over fit() method. https://arxiv.org/abs/2002.04688
19. Rusia J, Naugarhiya A, Majumder S, Majumdar S, Acharya B, Verma S (2016) RF based wireless data transmission between two FPGAs. In: 2016 International conference on ICT in business industry & government (ICTBIG), pp 1–6. https://doi.org/10.1109/ICTBIG.2016.7892643
20. Das M, Ansari W, Basak R (2020) Covid-19 face mask detection using tensorFlow, Keras and OpenCV. In: 2020 IEEE 17th India council international conference (INDICON), New Delhi, India, pp 1–5, https://doi.org/10.1109/INDICON49873.2020.9342585

Check for updates

Precision Agricultural Management Information Systems (PAMIS)

V. Lakshmi Narasimhan(✉)

Department of Computer Science, University of Botswana, Gaborone, Botswana
srikar1008@gmail.com

Abstract. Agriculture is still dominated by conventional plough and bull, while richer farmers employ tractors to plough both deep and wide. With the advent of the Internet of Things (IoTs) along with cheap sensors, one can engage in precision agriculture, wherein water and nutrients can be controlled over the field using drip water irrigation. This paper details the design of a Precision Agricultural Information Management Systems (PAMIS). The sensor network obtains dynamic information about soil and the use of Digital Elevation Models (DEMs) and emulation of Glasshouse can significantly aid precision agriculture. Cloud Computing services and long-term analysis and synthesis of large/Big datasets also facilitate precision agriculture. However, there are issues relating to privacy, security and legal, because of the use of Information Systems for precision agriculture. Further, readiness for entomological preventive studies can also be attempted, besides predicting harvest potentials within 3–4 weeks of seedling growth. The performance of the PAMIS Cloud has been evaluated using parametric modelling technique and the results indicate that PAMIS Cloud system can successfully enhance the performance of the scientists and technical people involved in this field; the details are also provided in this paper.

Keywords: Precision agriculture · Cloud computing · Internet of Things (IoT) · Sensor Networks and Precision Agricultural Management Information Systems (PAMIS) · Parametric performance modelling

1 Introduction

Long time ago, agriculture was dominated by conventional plough and bull and then came Tractors which can plough both deep and wide. Both the effectiveness and efficiency of agricultural operations were enhanced by these kinds of automation, which include other techniques such as automatic seedling planters. Later, drip water irrigation considerably optimized water and nutrients usage. Various types of timers encouraged Just-In-Time usage of various devices, thus enhancing overall agricultural production. Some kind of shading through implanted large trees and tarpaulins have also been used to climate control the fields in a limited fashion. The idea of intervening crop/plants along with the main crop in order to improve soil nutrition has always been employed.

Precision Agriculture relates to the use of Information and Communication Technology (ICT) for all agricultural related activities [1]. The advent of Internet of Things

R. Misra et al. (Eds.): ICIoTCT 2021, LNNS 340, pp. 162–174, 2022.
https://doi.org/10.1007/978-3-030-94507-7_16

(IoT) has made a quantum leap in ICT in many areas including medical, health, agriculture, aged-care and telemedicine, to name a few [2]. Simply put, IoTs have storage, processing power, sensing and communication power [11]. IoTs can also cooperate to work together, thereby creating a Cooperative Intelligent Agent system. Therefore they can sense, monitor and control various aspects of a given field, thereby seeding and solving issues in precision agriculture. However, precision agriculture does involve upfront investment for automating the agricultural field using IoT and related communications technology [3].

This paper details the design of a Precision Agricultural Information Management Systems (PAMIS). The rest of the paper is organized as follows: Sect. 2 introduces Precision Agriculture followed by Internet of Things (IoT) in Sect. 3 and the use of IoTs for Precision Agriculture using a network of IoTs and use of sensor network for obtaining dynamic information about soil in Sect. 4. Sections 5 and 6 respectively detail the use of Digital Elevation Models (DEMs) and points to ponder in the deployment of such systems. Section 7 presents a list of practical issues in field farming, while Sect. 8 explains the means to exploiting long-term analysis and synthesis of large/Big datasets. Section 9 outlines the issues relating to privacy, security and legal, because of the use of Information Systems for precision agriculture, while Sect. 10 indicates the readiness for entomological preventive studies. Section 11 provides the way by which one can predict harvest potentials within 3–4 weeks of seedling growth, followed by the description of the design of the PAMIS information architecture in Sect. 12. Section 13 details the parametric modeling technique used for the performance evaluation of PAMIS, while the Conclusions summarizes the paper and provides pointers for further work in this arena.

2 What is Precision Agriculture

The term Precision Agriculture relates to the large-scale deployment of sensors and communication devices over several inter-connected and cooperating Internet of Things (IoT) so that information about the soil, plants and the atmosphere can be collected and used for better agriculture production. Several types of datasets can be collected on a periodic basis from the IoTs using a multitude of sensors and the data collected can be analyzed using a variety of algorithms in order to predict a variety of factors, such as even output per acre of agricultural field in a given area [4]. However, IoTs need to be deployed properly and the data coming out of them can be affected by a variety of environmental factors, including dust, dirt, insects and others. Smart agriculture calls for systematic identification of all issues and usages and, ways and means to handle the datasets in context.

2.1 Relevance to Botswana

Botswana is not necessarily an agrarian economy, as 85% of Botswana is part of the Kalahari Desert. Botswana's agriculture is limited to the Okavango Peninsula, which is rich, but few farmers there follow principles of precision agriculture. Further, the nature of produces differs from one part to another and hence precision agriculture must be carefully introduced in this area. However, if Precision Agriculture principles are followed, productivity can increase manifold.

3 Internet of Things (IoT)

The Internet of Things (IoT) is nothing but a computer system (embedded and/or non-embedded) which can also be a distributed heterogeneous system. It can also be a real-time system, which can be Internet Enabled with Wired and/or Wireless communication. Further, an IoT can communicate bi-directionally and Geo-Aware. In addition, IoTs can take decisions on their own and/or as a team/network of IoTs. As a consequence, the set of IoT devices placed at strategic places can become a powerful monitoring system. Figure 1 captures a possible classification of IoT based systems; examples of IoT systems include (see Fig. 2) Heart Pacemakers and Household products, such as Microwave oven, Washer, Dryer.

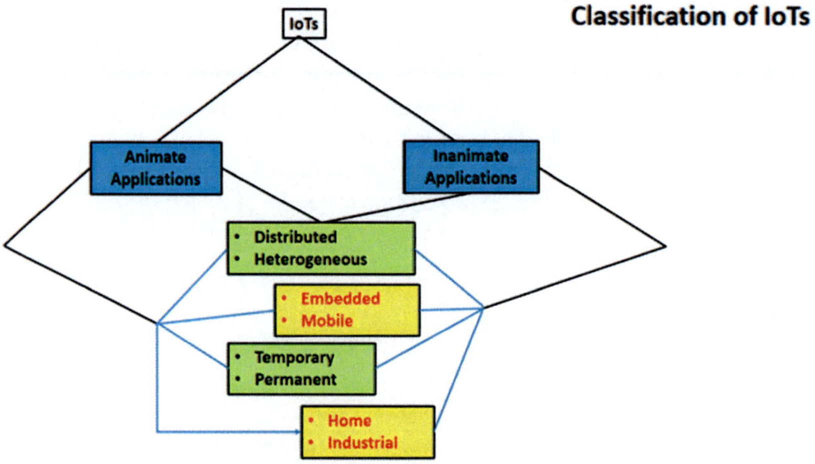

Fig. 1. Classification of IoT based systems

4 IoT for Precision Agriculture

IoTs for agriculture usually have several sensors that measure a variety of parameters such as: Temperature, pH, Moisture level, saturated vapor pressure, degree of Trace minerals (e.g., Phosphate, Potassium, Nitrogen, Oxygen, etc.). In addition, Agricultural IoTs can contain in-plant sensors that measure some the parameters (e.g., level of nutrients inside a plant) and Pressure sensors which can measure the flow of nutrients inside a plant. One can now create a sensor network which collects dynamic information about a given soil and also about a given plant type. Typical IoTs can communicate with each

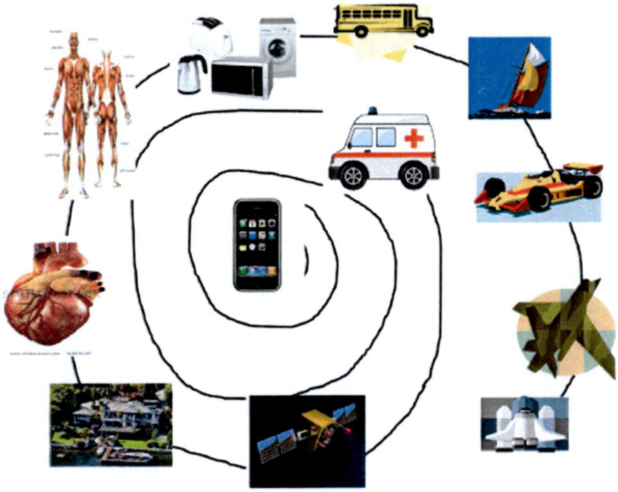

Fig. 2. Examples of IoT based systems

other and contain multiple sensors[1]. The sensors can form independent and autonomous units which can form dynamic networks of their own. As a consequence, dynamic information about the soil condition around the entire agricultural field can be obtained at the control station through several statistical algorithmic processes (e.g., averaging and normalization).

5 Using Digital Elevation Models (DEMs)

A 3-Dimensional Digital Elevation Model (DEM) [5] provides a better perspective of the landscape, particularly when the agricultural field is undulated as shown in Fig. 3. The DEM model also provide better understanding of the underlying hydrological considerations in the field. The DEM model will lead to the identification of areas where water or nutrients are logging and where they are in short supply, thereby resulting in better accuracy in soil sampling and soil classification. In addition, this will also lead to better cultivation planning and better crop siting. An understanding of the DEM model will also lead to better location of agricultural surveillance towers. There are software tools that can generate DEM model for a given field [5] and such software can be compared considering four aspects: i) price, ii) accuracy, iii) sampling density and, iv) pre-processing requirements.

6 Practical Points to Ponder

Points to ponder on IoT deployments in agricultural fields include, but not limited to, the following:

[1] A very interesting medical application with multiple sensors collecting information over a cloud environment is described in [10].

I. What are the best placements of IoT devices over a given field?
II. What are the vulnerabilities of IoTs in an agricultural environment?
III. How does one simplify generation of DEMs, given the technological/knowledge limitations of typical remote farmers?
IV. How would one employ Cloud Computing services for Precision Agriculture?
V. How would one go about exploiting long-term analysis and synthesis of large/Big datasets? How can one learn the long-term perspectives of Botswana's agricultural sector?
VI. How can the individual farmers' agricultural datasets be protected for the points of view of privacy, security and legal?
VII. What are the possibilities of hacking the datasets and, the creation of wrong and misleading datasets – particularly given the marketing forecast potentials of agricultural produces? (see Fig. 4 on the vast nature of datasets that can be collected)
VIII. To what extent the IoT-enabled field be readied for entomological preventive studies [6]?
IX. Can the harvest potentials of a given crop be predicted within 3–4 weeks of seedling growth?
X. What would be the overall information architecture underlying protocols for communication and data exchange between devices and sub-systems? A possible information architecture for this application is provided in Fig. 6.
XI. What would be the kind of Big Data algorithms and their applications to precision agriculture?
XII. How does one enhance the overall performance of the PAMIS system [12]?

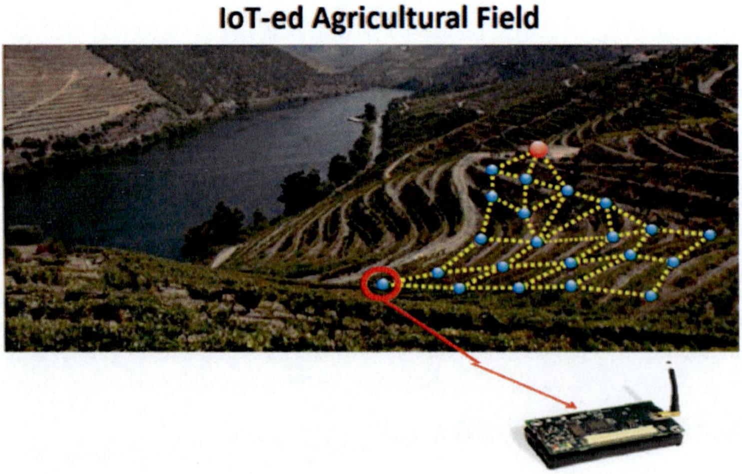

Fig. 3. IoTed agricultural field (adapted from [23])

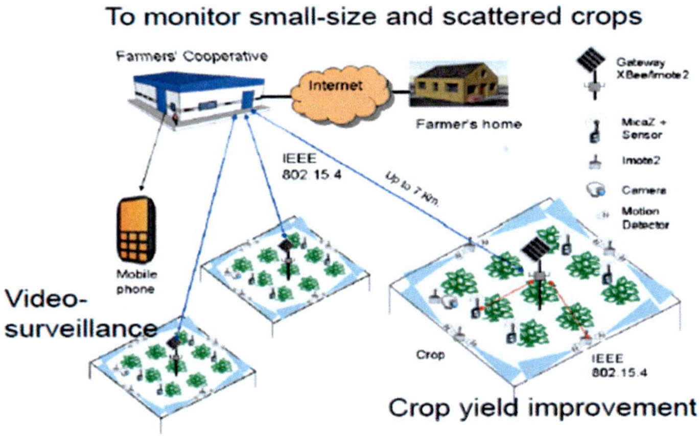

Fig. 4. Effecting improvements in crop yield (adapted from [23])

7 Practical Issues in Field Farming

IoT deployments in agricultural fields are fraught with practical difficulties which include, but not limited to, the following:

- Mud building over sensors and other parts of IoT thereby rusting them out.
- Animals, humans and equipment stomping over sensors and IoTs.
- Insects and birds building nests over sensors, thereby affecting the accuracy and resolution of sensors.
- Agricultural chemicals affecting various parts of IoT, particularly sensors and IoTs.
- Environmental decay of various parts of IoT, particularly sensors.

8 Exploiting Long-Term Analysis and Synthesis of Big Datasets

Long term analysis of datasets include, but not limited to, the following:

- Average nutrient level in the ground for a given crop type.
- Average moisture content in the ground.
- Crop inbreeding options and opportunities.
- Long-term In-plant sensor parameters that indicate average fluid flow inside plants in a given terrain.
- Predictive, Prescriptive and Pre-emptive analytics on various issues of agriculture – e.g., performance, pricing, etc.

9 Privacy, Security and Legal Issues

Issues relating to privacy, security and legal are many [8, 15], but the critical ones that affect farmers include the following:

Privacy Issues

- Would individual farmer's datasets be protected from "Agricultural Vulture Capitalist?"
- Would pricing of crop & their insurance values become vulnerable?

Security Issues

- What kind of access controls may be needed for storing, retrieving and protecting big datasets?
- Who or what will control data granularity and data Provenance?

Legal Issues

- Can Financial Institutions and Insurance companies demand such short-term and long-term datasets for their analyses?
- Are these datasets be demanded by the court of law?

10 Readying for Entomological Preventive Studies

It is well-known that insects typically follow wind patterns – actually most insects go against them as pollen is brought in by wind, while some insects help in pollination – but others can be pests. Advanced pest management is possible by combining wind patterns with agricultural data sets for both preventative and pollination studies. This approach calls for new and novel ways of acquiring data sets so that flowering and fruiting of plants along with wind patterns can be analyzed for entomologically active preventive programs [6].

11 Predicting Harvest Potentials Within 3–4 Weeks of Seedling Growth

The availability of agricultural big datasets along with crop models[2] can lead to predicting the amount of harvest ahead of time. "*Can the degree of harvest be predicted within 3–4 weeks of seedling growth?*" would be an excellent question to analyze, as this can relate to average price per quintal at harvest time and could potentially decide the time of planting and harvesting.

[2] Crop models along with soil models can provide a good prediction of the growth of a given crop – regarding rate of growth, flowering and fruiting timings, level of yield, etc.

12 Cloud Based Information Architectural Confluence of PAMIS

The architecture of PAMIS is mobile Cloud based [9, 18] with several services (see Fig. 5): i) Data as a Service (DaaS) [16] providing services for variety of Sensor data handling, IoT handling, Sensor life-cycle handling and Sensor provenance handling, ii) Software as a Service (SaaS) [17, 21] providing services for Feature location and extraction, DEM generation, Entomological data analysis, Data comparison, sub-image extraction, Image query processing, Historical analysis and Life-cycle analyses, iii) Portability & Interoperability as a Service (P-IaaS) [19, 22] providing services for Format conversion, Metadata Management, Data dictionaries, Glossary management, Data and information exporting and Interoperability Standards management, iv) Human Computing Interface as a Service (HCaaS) providing services for Decision support, Knowledge query management and HCI management, and, v) Generic Services providing services for Research data analytics, (generic) decision support, Cloud security management, Performance tuning, Office Services and Cloud help line. The architecture employs a confluence of technologies and communication protocols. The details of the architecture, along with its stage-wise optimizer will be covered in a later paper.

13 Parametric Performance Evaluation of the PAMIS Cloud

A parametric model based evaluation of the EDAM Cloud system has been carried out. Tables 1 and Table 2 provides typical parameters used for the evaluation of the EDAM Cloud, which have been obtained after discussions with several experts. Table 3 provides a list of performance indicators and their values, wherein the values are calculated using Relative Cost Unit (RCU) so that depending on the actual cost of individual companies a suitable multiplier can be employed to calculate the corrected values of various performance indices. It is hoped that these indicators will provide the way forward for the advancement of such systems in various marine sensor network R&D centers around the world.

Table 1. Parameters for evaluating PAMIS-cloud information systems architecture

S. no.	Explanation	Symbol	Average value	Max value
1	Size of PDF Agri-file	a	0.001 GB	1 GB
2	Number of sensors per acre	b	10	40
3	Number of acres per field	c	20	35
4	Average number of messages per sensor	d	3	20
5	Number of gateway nodes	e	5	7
6	Average sensor message size per message	f	5	10
7	Number of services in PAMIS	g	4	10
8	Number of maintenance calls per day	h	4	6
9	Number of *To-Act-On* messages per day	i	5	7

(continued)

Table 1. (*continued*)

S. no.	Explanation	Symbol	Average value	Max value
10	Number of upgrade requirements per day	j	4	6
11	Number of internal low-end service calls per day	k	10	18
12	Number of internal medium-end service calls per day	l	5	8
13	Number of internal high-end service calls per day	m	3	5
14	Number of reports to be generated per day	n	40	60
15	Number of knowledge query management per day	p	50	70
16	Number of help line management--simple call per day	q	20	30
17	Number of help line management--medium call per day	r	10	15
18	Number of help line management--complex call per day	s	5	8
19	Number of compliance requirements per day (if any)	t	1	3
20	Average viewing time per report	u	4 min	10 min

Table 2. Parameters for evaluating PAMIS-cloud information systems architecture

S. no.	Explanation	Symbol	Relative cost units (RCU)
1	Data storage cost per GB per month	C1	25
2	Data access cost per GB	C2	2
3	Internal low-end service cost per service call	C3	1
4	Internal medium-end service cost per service call	C4	3
5	Internal high-end service cost per service call	C5	8
6	Maintenance cost per service call	C6	15
7	Upgrade cost per service call	C7	10
8	Encryption cost per file (1 MB)	C8	5
9	Decryption cost per file (1 MB)	C9	5
10	Air conditioning costs per day	C10	200
11	Average downtown costs for services upgrade per day	C11	400

(*continued*)

Table 2. *(continued)*

S. no.	Explanation	Symbol	Relative cost units (RCU)
12	Compliance management costs per compliance requirement (if any)	C12	500
13	Average report generation cost per report	C13	10
14	Average knowledge query management cost per query	C14	2
15	Help line management per simple call	C15	1
16	Help line management per medium call	C16	5
17	Help line management per complex call	C17	10

Table 3. PAMIS--precision agriculture cloud performance indicators

S. no.	Metric name	Symbol	Formula	Typical average value	Max. value
1	Average execution time per acre	PI-1	(a * b + d * e * g) * h*i	1,500.2	60,480
2	Average bandwidth used per day	PI-2	a * b * c * i	1	9,800
3	Average downtime management per acre	PI-3	C11/ (b * c)	2	0.29
4	Average cost of security per acre	PI-4	(C8 + C9) * b * e * d	1,500	56,000
5	Average ease of use per acre = = Average execution time per acre + Weighted average service call time + Weighted average Help call time	PI-5	{(a * b + d * e * g) * h*I} + {C3 * k + C4 * l + C5 * m} + {C15 * q + C16 * r + C17 *s}	1,669.2	60,747
6	Average report generation cost per day	PI-6	C13 * n	400	600
7	Average compliance requirement cost per day (if any)	PI-7	C12 * t	500	1,500

(continued)

Table 3. (*continued*)

S. no.	Metric name	Symbol	Formula	Typical average value	Max. value
8	Average network usage cost = = Average Execution Time acre cost + Visit cost + Specialty related cost + InfoSec cost + Data access & storage cost	PI-8	(a * b * c) + (d * f) + (g * i) + (C8 + C9) * i + (C1 + C2) * b * c * i	27,085.2	266,340
9	Average PAMIS cloud usage cost = = PI-18 + Knowledge query cost + upgrade cost + maintenance cost + Aircon cost	PI-9	PI-8 + (C14 * p) + (C7 *j) + (C6 * h) + C10	27,485.2	267,170
10	Average cost of ownership per acre	PI-10	PI-9/c	1,374.26	7,633.43

14 Conclusions

Precision Agriculture is now the future and it can be cost-effective even in Botswana. The Internet of Things (IoT) is useful for precision agriculture. The underlying sensor networks can collect dynamic information about soil and store them. In addition, Digital Elevation Models (DEMs) & Glasshouse based DEMs & Emulation are useful for Precision Agriculture, besides the fact that Cloud Computing services are a must for Precision Agriculture. Exploiting long-term analysis and synthesis of large/Big datasets would be useful for Precision Agriculture – including readying agricultural field for entomological preventive studies. Privacy, security and legal issues are important issues to consider in Precision Agriculture. Furthermore, predicting harvest potentials within 3–4 weeks of seedling growth is now possible so that price per quintal can be predicted (and hence the profits thereof); appropriate storage mechanism and related constraints therein can also devised. Global optimization of various parameters is also now possible, leading to better overall agricultural management, performance and profit margins. Future research in this arena include: i) preventing sensor degradation or computationally compensating thereof, ii) protecting field IoTs, iii) developing newer data analytics/algorithms for better precision agriculture and iv) enhancing entomological studies, particularly for horticulture applications. The performance of the PAMIS Cloud has been evaluated using parametric modelling technique and the results indicate that PAMIS Cloud system can successfully enhance the performance of the scientists and technical people involved in this field; the details are also provided in this paper.

References

1. Precision Agriculture: https://whatis.techtarget.com/definition/precision-agriculture-precision-farming. Accessed 15 Sept 2019
2. Internet of Things (IoT): https://en.wikipedia.org/wiki/Internet_of_things. Accessed 15 Sept 2019
3. Benefits and Costs of Entry Level Precision Agriculture: https://nydairyadmin.cce.cornell.edu/uploads/doc_410.pdf. Accessed 15 Sept 2019
4. Jones JW et al (2003) The DSSAT cropping system model. Eur J Agron 18(3–4):235–265
5. NASA World Wind: Opensource GIS for Mission Operations (DEM Model Development): https://ieeexplore.ieee.org/abstract/document/4161692. Accessed 15 Sept 2019
6. Imms AD (2012) Recent advances in entomology. London
7. NIST Definition of Cloud Computing. https://csrc.nist.gov/publications/detail/sp/800-145/final. Accessed 14 Sept 2019
8. NIST Cloud Computing Security Reference Architecture. https://csrc.nist.gov/publications/detail/book/2016/cloud-computing-security-essentials-and-architecture. Accessed 14 Sept 2019
9. De D (2016) Mobile cloud computing: architectures, algorithms and applications. CRC Press, Boca Raton
10. Lakshmi Narasimhan V (2019) Botswana's lab-ın-A-briefcase – a position paper. In: ACM press proceedings of the Australasian computer science week (ACSW 2019), Sydney, Australia, pp 29–31, January 2019
11. Arduino. https://www.arduino.cc/. Accessed 14 Sept 2019
12. Lakshmi Narasimhan V, Jithin VS (2018) Time-cost effective algorithms for cloud workflow scheduling - extension of an earlier work. In: Proceedings of IST Africa ınternational conference Gaborone, Botswana, 9–11 May 2018. IEEE Xplore
13. India's Department of Electronics and Information Technology (DeitY). www.deity.gov.in. Accessed 14 Sept 2019
14. India's Ministry of Electronics and Information Technology Initiative (MeitY). www.meity.gov.in. Accessed 14 Sept 2019
15. Lakshmi Narasimhan V (2013) Research issues and challenges in cloud computing- a critical perspective. In: National seminar on computational ıntelligence (NSCI 2013), SRM University, Kattangalattur, India, 21–22, January 2013
16. Data as a service: https://searchdatamanagement.techtarget.com/definition/data-as-a-service. Accessed 5 June 2019
17. Software as a service: https://searchcloudcomputing.techtarget.com/definition/Software-as-a-Service Accessed 5 June 2019
18. Intel Cloud Computing: Transform IT for a Hyper Connected World – Intel Cloud Computing. https://www.intel.com/content/www/us/en/cloud-computing/overview.html. Accessed 17 Sept 2019
19. Infrastructure as a Service – Tech Target Infrastructure as a Service – Tech Target. https://searchcloudcomputing.techtarget.com/definition/Infrastructure-as-a-Service-IaaS. Accessed 17 Sept 2019
20. Benefits of Cloud Computing in Business. https://www.eukhost.com/blog/webhosting/10-benefits-of-cloud-computing-for-businesses/. Accessed 17 Sept 2019

21. Software as a Service – Tech Target. https://searchcloudcomputing.techtarget.com/definition/Software-as-a-Service. Accessed 17 Sept 2019
22. IaaS vs DaaS vs PaaS vs SaaS – Which should you choose? – ESDS Blog. https://www.esds.co.in/blog/iaas-vs-daas-vs-paas-vs-saas-which-should-you-choose/. Accessed 17 Sept 2019
23. Le TD, Tan TH (2015) Design and deploy a wireless sensor network for precision agriculture. In: 2nd national foundation for science and technology development conference on information and computer science (NICS), https://doi.org/10.1109/nics.2015.7302210
24. Upscaling of greenhouse vegetable production: https://www.cbs.nl/en-gb/news/2018/16/upscaling-of-greenhouse-vegetable-production. Accessed 17 Sept 2019

Vision for Eyes

Jaya Rishita Pasam, Sai Ramya Kasumurthy, Likith Vishal Boddeda(✉), Vineela Mandava, and Vijay Varma Sana

Department of CSE, GITAM Institute of Technology, Visakhapatnam, Andhra Pradesh, India

Abstract. Spending too many hours in front of a monitor, screen or pc will definitely have a great impact on our eyes. There might be many products which make business in the market by advertising that their products reduce the strain on eyes like optimizing the blue light emission from the screen, and few other screen guards which will cause less harm to the eyes even if we expose our eyes for a long time to the screen. According to the American Academy of Ophthalmology, there exists no such product that actually reduces strain on the eyes. So we are a start-up building a device and service which can manually decrease the strain on the eyes and the human brain which can be caused due to extreme exposure to the display screens with high contrasts and glaze.

Keywords: Raspberry pi · Eyes · IoT · Object distance measurement · Facial recognition · Object detection · Eye strain · Eye protection

1 Introduction

We are a start-up team working on building a product which is a combination of both software and hardware. Basically, it is a small portable device that can be placed at the top of the screen, it can be removed whenever the user wants to. The functionality of the product is simple, it will remind the user when they are so close to the screen or monitor. But it will not remind me whenever an object is close to the screen, it will be functioning only if the object is a human face.

By capturing the face of the human, our device will recognize that it is a face and locate the position of the eyes, after the detection of the eyes we are adding a feature that will let the user know how much their eyes got strained. The product will be developed using IoT technology. The technologies that we would possibly use to build the product are as follows:

IoT:

- Raspberry Pi [python]
- Infrared Human Sensor or HC SR04 Ultrasonic Sensor
- LED, male to the female jumper, breadboard

Computer Vision:

- OpenCV [python]
- Image Processing

R. Misra et al. (Eds.): ICIoTCT 2021, LNNS 340, pp. 175–184, 2022.
https://doi.org/10.1007/978-3-030-94507-7_17

2 Literature Survey

We first analyzed our approach towards product development. We found that there has been a problem for many years. So we then realized what aspects and markets we should concentrate on. Here comes the problem statement, we all know that the present IT Industry demands continuous sitting and working in front of laptops/PCs which might cause some serious health-related problems. Directly or indirectly the user is getting affected, the user may or may not experience the effect soon, but in the long run, they will have to face it and they are facing it too. So how do we overcome this? We can manually take care of our eyes and control the strain by following the 20-20-20 rule. It states that for every 20 min we need to take a break for 20 s and look at an object which is 20 feet away from us. This keeps our eyes focused, which can really stimulate. There are other precautions to be followed to prevent eye strain, we need to continuously blink our eyes for 20 to 30 s once an hour or 30 min. But most people do not do this. Here comes our product which will remind the user when their eyes are getting affected directly or indirectly. We then designed a rough sketch of our product, by mentioning all the equipment to be used. With a thought of the end product, we designed a rough sketch of it.

It is simple when a user gets close to the screen, our product would remind them to move a little backward to make sure that the glaze emitted from the screen will not affect the eyes (Fig. 1).

Fig. 1. Pictorial view of the model

The application of the product is so simple, first, we set up the raspberry pi model and program it so that when the camera recognizes a face near to the screen (the distance is measured with help of HC-sensor) the LED will flash, as the user moves back, the sensor will calculate the distance and acknowledge whether the user is close to the screen or not. Components we need are, a raspberry pi, HC-sensor, breadboard, LED light, Envirochip, male to female jumpers, and a camera module if you are not using the camera of your laptop or PC.

3 Problem Identification and Objectives

- The present IT Industry demands continuous sitting and working in front of the laptops/PCs which might cause some serious health-related problems.
- A few of the reasons being staring at the screen continuously for way too long, not maintaining a certain distance from the monitor, radiation emitted etc.
- The serious health problems that may occur due to excessive use of computers include migraine, myopia, impact on one's sleep, body tissues, etc.

4 System Methodology

To build an IoT based project that can detect the strain of an eye, the following steps are to be accomplished:

1. Establish the IoT raspberry pi setup with HC sensor and computer vision software.
2. Program the raspberry pi model to detect the face and calculate the distance.
3. After capturing the face, if the user is close to the screen i.e. less than the threshold that is to be mentioned, the system alerts the user to move back since their eyes are getting strained.

5 Implementation

5.1 Hardware Requirement

- Raspberry pi model 4
- RAM 4 GB
- Minimum 8 GB memory card
- Camera module
- Two 1 Ω resistors
- LED
- HC-SENSOR
- Male to Female jumpers
- Breadboard
- Monitor or a screen
- Keyboard and Mouse

5.2 Software Requirements

- Raspberry pi OS
- RPi.GPIO packages
- OpenCV

5.3 Setup

- Boot the raspberry pi device with Raspberry pi OS
- install open_cv for python on the device via command line or download the package
- Also, install RPi.GPIO package to deal with the GPIO pins

After installing all the packages, we are ready to implement our idea. Now that we have all the hardware equipment, we need to assemble them and connect them to the raspberry pi board.

Firstly, connect the HC-Sensor to the breadboard on any of the row (i.e. A,B,C,D,E,F,G,H,I,J). After connecting the HC-Sensor, we need to connect its 4 pins i.e. Vcc, Trig, Echo, GND to the raspberry pi GPIO pins. In the proposed model we are connecting the Vcc of the HC-Sensor to 5v pin in the pi, Trig pin to the GPIO4. Now the Echo of the HC-Sensor i.e. signal input pin should be connected by a 1-ohm resistor and connecting to the GPIO18, GND pin to be connected to the ground of the pi board and also connect a 1-ohm resistor from the GND pin in parallel to the resistor of Echo pin. All the interfacing between raspberry pi and breadboard is carried out by male-to-female jumpers. Now we have successfully configured the HC-Sensor and interfaced it with the raspberry pi board. In order to alert the user, we are going to use a led and flash it when the user gets closer to the screen. To perform the activity we need to first connect the led to the raspberry pi device. Firstly establish a ground connection to the board at the terminal end in order to supply power to the LED. Place a LED on any row on the breadboard, connect a male-to-female jumper at any positive terminal on the breadboard with any GND pin in the pi, connect the one end of the 1-ohm resistor in parallel to the connected GND pin on the positive terminal on the breadboard and the other end at the negative terminal of the LED. Now connect a male-to-female jumper to the positive terminal of the LED and interface it with GPIO17 which can be used as an output signal. We can now consider that whenever the signal is high at GPIO17, the LED will glow. After all the interfacing is done, now connect the camera module directly to the raspberry pi board (Fig. 2 and 3).

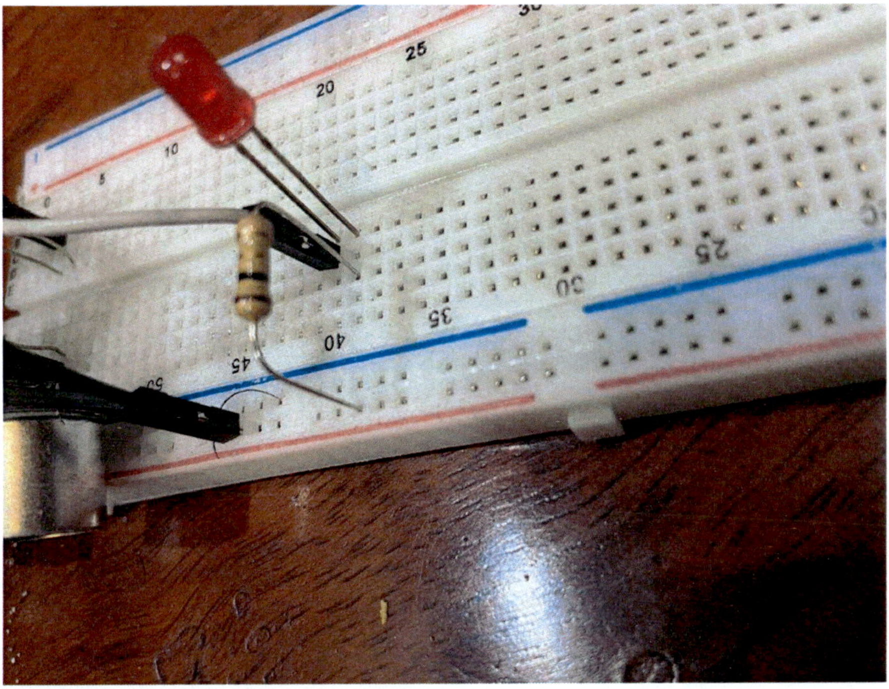

Fig. 2. LED light connection on a breadboard to raspberry pi

5.4 Execution Methodology

The program task is to first calculate the distance from the device, if the distance is less than a threshold value then it will look out to detect any human faces.

We will first initialize the GPIO pin values for TRIG and ECHO and then set up the GPIO17 pin to output the signal. Now we define a function to calculate the distance of the object which may appear in front of the device and we return the distance from the function. Later after we compute the distance, we will then detect any faces on the object which is closer to the sensor and then alert the user accordingly. This is a real-time implementation, so we choose to capture the movements in real-time by accessing the camera module (Fig. 4).

Below is the code snippet to check the threshold value which is mentioned to limit the distance of the object from the device and if the distance is less than 20 cm then the user will get an alert by flashing the LED light (Fig. 5, 6 and 7).

Fig. 3. HC-SENSOR connection

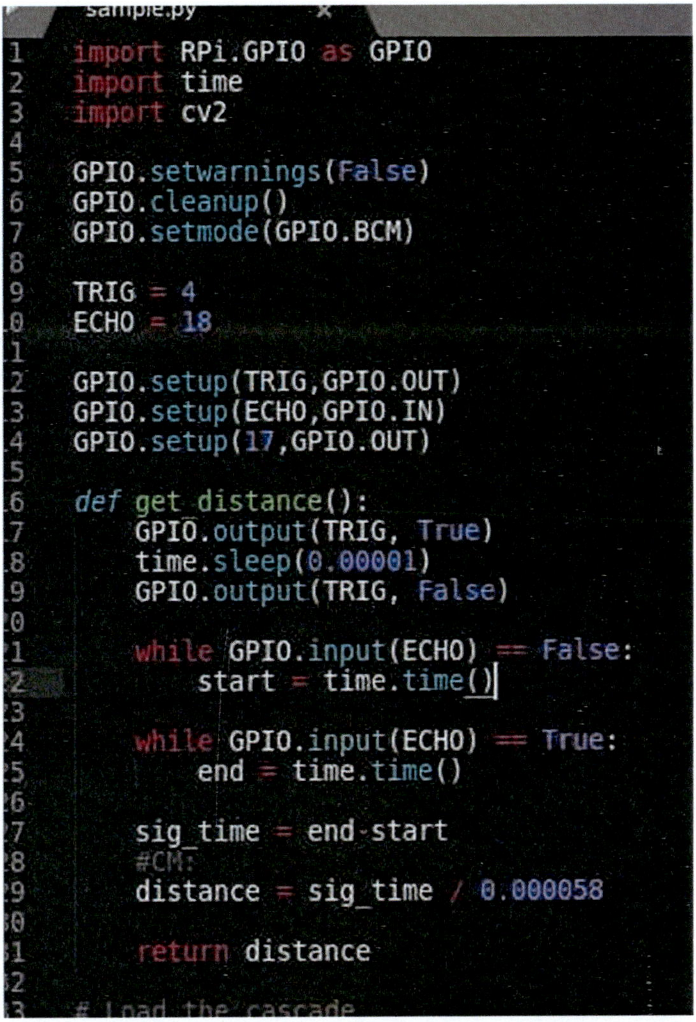

```
import RPi.GPIO as GPIO
import time
import cv2

GPIO.setwarnings(False)
GPIO.cleanup()
GPIO.setmode(GPIO.BCM)

TRIG = 4
ECHO = 18

GPIO.setup(TRIG,GPIO.OUT)
GPIO.setup(ECHO,GPIO.IN)
GPIO.setup(17,GPIO.OUT)

def get_distance():
    GPIO.output(TRIG, True)
    time.sleep(0.00001)
    GPIO.output(TRIG, False)

    while GPIO.input(ECHO) == False:
        start = time.time()

    while GPIO.input(ECHO) == True:
        end = time.time()

    sig_time = end-start
    #CM:
    distance = sig_time / 0.000058

    return distance

# Load the cascade
```

Fig. 4. Code snippet to return the distance of an object.

```
distance = get_distance()

#inches:
#distance = sig_time / 0.000148
if distance > 20:
    print('Distance: {} centimeters'.format(distance))
elif value == 1 and distance < 20:

    GPIO.output(17,GPIO.HIGH)
    time.sleep(3)
    GPIO.output(17,GPIO.LOW)
    print("move back your distance is {}".format(distance))

elif value == 0 and distance < 20:
    print('Distance: {} centimeters'.format(distance))
```

Fig. 5. Code snippet to check the distance

```
# Load the cascade
face_cascade = cv2.CascadeClassifier('haarcascade_frontalface_default.xml')

# To capture video from webcam.
cap = cv2.VideoCapture(0)
# To use a video file as input
#cap = cv2.VideoCapture('selfvid.mp4')

while True:
    # Read the frame
    _, img = cap.read()

    # Convert to grayscale
    gray = cv2.cvtColor(img, cv2.COLOR_BGR2GRAY)

    # Detect the faces
    faces = face_cascade.detectMultiScale(gray, 1.1, 4)

    # Draw the rectangle around each face
    for (x, y, w, h) in faces:
        cv2.rectangle(img, (x, y), (x+w, y+h), (255, 0, 0), 2)

    # Display
    cv2.imshow('img', img)

    #print ("Found {}" + str(len(faces)) + " face(s)")
    value = len(faces)

    # Stop if escape key is pressed

    distance = get_distance()
```

Fig. 6. Code snippet to capture the face and return 1 if found a face

Fig. 7. Program output

6 Testing

In the above figure, we can see multiple statements with distance and the count of the faces. They are totally three type of cases solved and prompted on the screen and they are as follows:

- When detected faces but the face object is not below the distance threshold, so there will be no alert and just return the distance of the object from the device
- When no faces are detected, it returns only the distance of the object from the device.
- when faces are detected and they are below the threshold value, the users get an alert and the led will flash.

7 Results

By the implementation of the above code in the raspberry pi model 3, we can observe that by capturing the image and calculating the distance between the object and screen, the program checks whether these conditions are true or false and then alerts the true.

8 Conclusion

In this report, we demonstrated a method to alert the user in front of a monitor or a screen by flashing a LED light if the user is close to the screen, using the OpenCV and Raspberry pi. To the best of our knowledge, this is the first effective way that alerts a user when their eyes are getting affected.

References

1. Kanade P, Alva P (2020) Raspberry PI project-ultrasonic distance sensor in civil engineering, pp 2321–1776. https://doi.org/10.5281/zenodo.4392971

2. Tyas Purwa Hapsari D, Gusti Berlina C, Winda P, Arief Soeleman M (2018) Face detection using Haar cascade in difference illumination. In: 2018 international seminar on application for technology of information and communication, pp 555–559. https://doi.org/10.1109/ISEMANTIC.2018.8549752
3. Likith Vishal B et al (2020) Image classification using neural networks and tensor-flow. In: Test engineering and management, vol 83, pp 20087–20091, ISSN 0193-4120, March–April 2020
4. Maksimovic M, Vujovic V, Davidović N, Milosevic V, Perisic B (2014) Raspberry Pi as Internet of Things hardware: Performances and Constraints
5. Parveen S, Shah J (2021) A motion detection system in python and Opencv. In: 2021 third international conference on intelligent communication technologies and virtual mobile networks (ICICV), pp 1378–1382. https://doi.org/10.1109/ICICV50876.2021.9388404

Wheat Disease Severity Estimation: A Deep Learning Approach

Sapna Nigam[1], Rajni Jain[2](✉), Surya Prakash[3], Sudeep Marwaha[1], Alka Arora[1], Vaibhav Kumar Singh[4], Avesh Kumar Singh[5], and T. L. Prakasha[6]

[1] ICAR-Indian Agricultural Statistics Research Institute, New Delhi, India
sapna.nigam@icar.gov.in
[2] ICAR-National Institute of Agricultural Economics and Policy Research, New Delhi, India
rajni.jain@icar.gov.in
[3] ICAR-Indian Institute of Technology, Indore, India
[4] ICAR-Indian Agricultural Research Institute, New Delhi, India
[5] Punjab Agricultural University, Ludhiana, Punjab, India
[6] ICAR-Indian Agricultural Research Institute Regional Station, Indore, Madhya Pradesh, India

Abstract. In the agriculture domain, automatic and accurate estimation of disease severity in plants is a very challenging research field and most crucial for disease management, crop yield loss prediction and world food security. Deep learning, the latest breakthrough in artificial intelligence era, is promising for fine-grained plant disease severity classification, as it avoids manual feature extraction and labor-intensive segmentation. In this work, the authors have developed a deep learning model for evaluating the image-based stem rust disease severity in wheat crop. Real-life experimental field conditions were considered by the authors for the image dataset collection. The stem rust severity is further classified into four different severity stages named as healthy stage, early stage, middle stage, and end-stage. A deep learning model based on convolutional neural network architecture is developed to estimate the severity of the disease from the images. The training and testing accuracy of the model reached 98.41% and 96.42% respectively. This proposed model may have a great potential in stem rust severity estimation with higher accuracy and much less computational cost. The experimental results demonstrate the utility and efficiency of the network.

Keywords: Deep learning · Image classification · Plant disease severity · Wheat rust

1 Introduction

Plant diseases pose a significant threat to agricultural production losses. Disease identification and severity is a major concern for farmers in terms of reduced crop production as well as the crop yield loss. This leads to economic loss and food insecurity in many areas. Plant disease severity is a critical parameter for determining the disease level and as a result, can be used to forecast yield and recommend control measures. Hence, preventive action is required for the disease identification in plants at an early stage. The ability to

R. Misra et al. (Eds.): ICIoTCT 2021, LNNS 340, pp. 185–193, 2022.
https://doi.org/10.1007/978-3-030-94507-7_18

diagnose disease severity quickly and accurately would aid in reducing yield losses [1]. Plant disease severity is traditionally determined by domain experts visually inspecting the plant tissues. Modern agriculture's rapid growth is stymied by the high cost and low efficiency of human disease evaluation [2]. Precision farming, high-throughput plant phenotype, and other fields are striving for automated disease diagnosis models, due to the advent of digital cameras and advances in computer vision. To overcome this concern, automated systems and techniques for detecting plant diseases are needed that will take less time and effort with much higher accuracy as compared to other conventional ways [3, 4].

[5] reviewed and studied the comparison of Deep Learning (DL) techniques which has gain momentum in recent years. Deep learning is when a neural network learns hierarchical data representations with several abstraction layers [6, 7]. Their results show that deep learning outperforms commonly used image processing techniques in terms of accuracy [8, 9]. While existing plant disease detection and diagnosis procedures are reliable, they are inadequate when it comes to assessing disease severity. [10] annotated the healthy and black rot images of apple in the Plant Village dataset with various severity categories. The Leaf Doctor app [11], an interactive smartphone app, may be used on colour images to distinguish lesion areas from healthy tissues and calculate disease severity percentages. This application even outperformed the Assess in terms of accuracy. A novel deep learning architecture named PD2SE-Net was constructed with accuracies of 91%, 98%, and 9%, respectively, for evaluating plant disease intensity and categorization, as well as plant species identification, using ResNet50 as the basic model [12]. There is considerable intraclass similarity and modest inter-class variation, fine-grained disease severity classification is substantially more challenging than classification among distinct diseases [10, 13]. Deep learning is ideal approach for fine-grained disease severity classification because it avoids time-consuming feature extraction and threshold-based segmentation [14].

This research was motivated by a breakthrough in deep learning for image-based plant disease recognition [15, 16]. It proposes a deep learning model for automated image-based diagnosis of plant disease severity in wheat caused by stem rust disease. Wheat rusts have been the most important biotic stresses responsible for unstable crop production. The wheat crop has three types of rusts named yellow rust, leaf rust, and stem rust. If not properly managed, these diseases can infect and cause significant yield losses. Rusts are known for spreading quickly and reducing wheat yield and quality. [17] identified the wheat yellow rust from the healthy leaves using deep learning. However, this research, on the other hand, is solely focused on stem rust and its severity estimation based on the different severity levels, such as Healthy Stage, Early Stage, Middle Stage, and End Stage. Therefore, the aim of this study is to develop a deep learning model for wheat stem rust disease severity estimation that will have the caliber to correctly predict the disease severity stage from the input image.

2 Methodology

The complete methodology for the experiment is summarized in Fig. 1. The images were collected from the experimental field according to the four different disease severity stages. In second step, the images were preprocessed. The detailed image preprocessing done is mentioned in Sect. 2.2. After the image preprocessing, the images were fed into the Convolution neural network (CNN) as an input for feature extraction and classification.

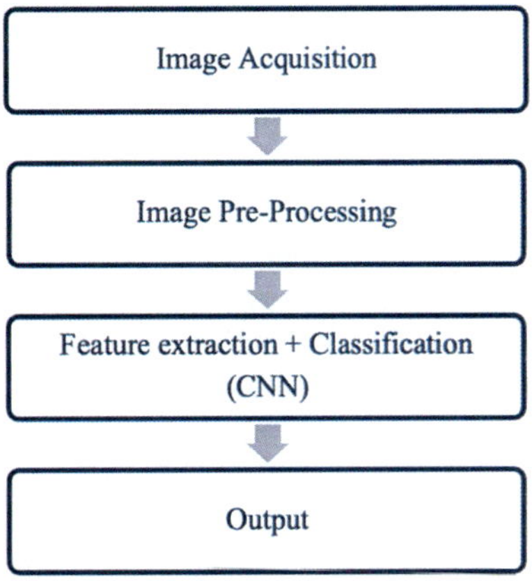

Fig. 1. Methodology for image-based classification

Convolution neural network architecture is one of the most promising deep learning architectures for image-based classification (CNN). The three layers that make up a CNN architecture are the convolution layer, pooling layer, and fully connected layer. Convolution Layer automatically extracts features from each input image. It is made up of a set of learnable filters that train the relationship between features and use kernel or filters to build a feature map. During training, CNN employs the Rectified Linear Unit (ReLU), which has an output of (x) = max (0, x) and introduces non-linearity into the network. The pooling layer downsamples convolution maps, decreasing training time and preventing overfitting by keeping just the most useful information for future processing. The final pooling layer output (3D matrix) is flattened into a one-dimensional vector, which is then used as the input in a fully connected layer. The features are then combined to form a model. Finally, the SoftMax or sigmoid activation function computes the predefined class scores and assigns the image to one of them. This study's Convolution neural network architecture (CNN) is shown below (Fig. 2).

The stem rust infected image is given input to the model. In the first phase of feature extraction, the features are automatically extracted from the images in convolution layers and down sampled in pooling layers. In the second phase of classification, the layers are first flattened and fully connected layers further classify the input image into four different classes as mentioned above.

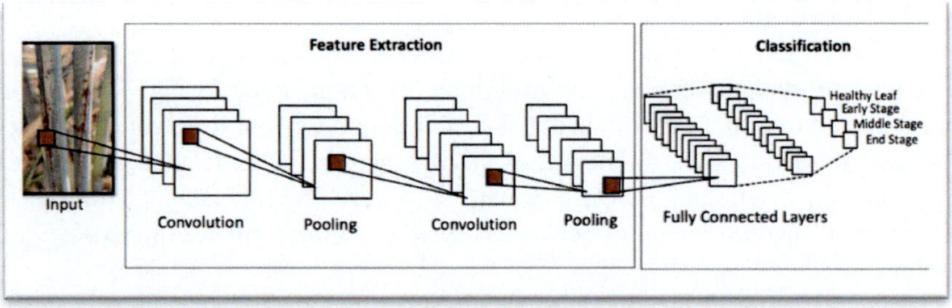

Fig. 2. CNN architecture for image-based plant severity estimation

2.1 Dataset

The author has created the dataset from the real-life conditions. The dataset consists of stem rust images collected from the experimental fields of ICAR-Indian Agricultural Research Institute Regional Station, Indore from January 2021 to March 2021. The pustules in stem rust are much larger, orange-red, oval to elongated, and appear on the stem, leaf blade, and sheath, as well as parts of the spike (Fig. 3). The images were collected keeping in mind the four levels of severity.

Healthy Stage Early Stage Middle Stage End-Stage

Fig. 3. The different four severity stages of Stem Rust disease

The author has divided these stages into four classes 0, 1, 2, and 3. These classes can also be referred to as the Severity scale for the stem rust disease. Table 1 provides a detailed description of the data.

Table 1 shows that there were a total of 2587 images collected, which were classified into four categories. The dataset is further split into a training set, testing set, and validation set into 80: 20:10 for the network input.

Table 1. The number of images in each severity stage.

Classes (severity scale)	Severity stage	Total no. of images (Train set + Test set)	Severity range (%)
0	Healthy stage	580 + 145 = 725	0
1	Early stage	624 + 156 = 780	0–25
2	Middle stage	564 + 142 = 706	25–50
3	End-stage	278 + 98 = 376	>50

2.2 Image Pre-processing

The author-created dataset mentioned in the previous section consists of arbitrarily sized RGB images. In deep learning models with effective end-to-end learning, only basic steps for image pre-processing are required. Images are processed in this stage according to these steps. For our network, firstly, all of the images to 256 * 256 pixels were resized. On these rescaled images, both model optimization and prediction were performed. Second, all pixel values are divided by 255 in order to match the network's initial values. In the third step, the training images are subjected to a variety of random augmentations such as rotation, shearing, and flipping. The augmentation helps the model generalize better by preventing over-fitting.

2.3 Implementation

The experiment for this study is performed on an Ubuntu workstation having Intel Xeon (R) Silver 4214 CPU (125.6 GB), accelerated by Quadro RTX 4000 graphics card. The Anaconda environment, with the Keras framework and the Tensorflow at the backend, is used to implement the model (Table 2).

Table 2. Software and hardware specifications used for the experiment

S. no.	Software and hardware specifications	
1	Operating system	Ubuntu
2	Workstation configuration	Intel Xeon (R) Silver 4214 CPU (125.6 GB)
3	Graphics card	Quadro RTX 4000
4	Environment	Anaconda
5	Framework	Keras with TensorFlow at the backend

2.4 Model Developed

Six convolutional layers, five Max Pooling layers, and two fully linked layers make up the neural network architecture. Previously, we downsized all of the photos for our network to 256 * 256 pixels, and rescaled images with a dimension of 256 * 256 were sent to the network. The size of the filter utilised is 3 * 3. To achieve non-linearity, we employed a rectified linear unit (ReLU) after the convolution layer.The first convolutional layer has 16 filters, the second has 20, and so on, with the number of filters increasing layer by layer. There are no paddings in any of the convolutional layers. We employed 128 neurons in the first completely linked layer. Before being supplied to a fully linked layer, the data is flattened. After each convolutional layer, a max-pooling layer decreases the dimensionality, and a 20% drop-out prevents the model from over-fitting. A sigmoid activation function is used in the last fully connected layer to generate probability distributions for each of the four classes. The batch size for training data is set to 20 and the epoch count was set to 40. The model compilation consists of the Adamax optimizer and Sparse Categorical Crossentropy loss, to handle the imbalanced number of pixels for each class. The learning rate was set to 0.001. The total number of trained parameters is 631,228. All the hyperparameters are mentioned in Table 3.

Table 3. Hyperparameters used in the experiment

S. no.	Hyperparameters	Value
1	Filter size	3 * 3
2	Batch size	20
3	Epochs	40
4	Optimizer	Adamax
5	Loss	Sparse categorical crossentropy
6	Learning rate	0.001
7	Dropout	20%
8	Padding	0

3 Results and Discussion

A batch size of 20 was set for the model, which was trained on 40 epochs. The training accuracy at 40 epochs is 98.41% and validation accuracy is 97.55%. The number of iterations per epoch is 93. We trained our model from scratch. The test data consist of 541 images from the different classes. In model testing, we observed that the overall average model testing accuracy was 96.42%. Therefore, it can conclude that model has the potential for performing real-time diagnosis for plant disease severity based on these four stages.

Therefore, in Fig. 4(a) and Fig. 4(b), it is clear to see from the above curves that the model has been well trained. It can conclude that it is a good fit model because training loss decreases to a point of stability and a slight difference between the train and validation loss learning curves is observed.

The confusion Matrix for four different classes is shown in Fig. 5. It can be observed that 5 healthy stage images out of 145 were misclassified as early-stage disease. Similarly,

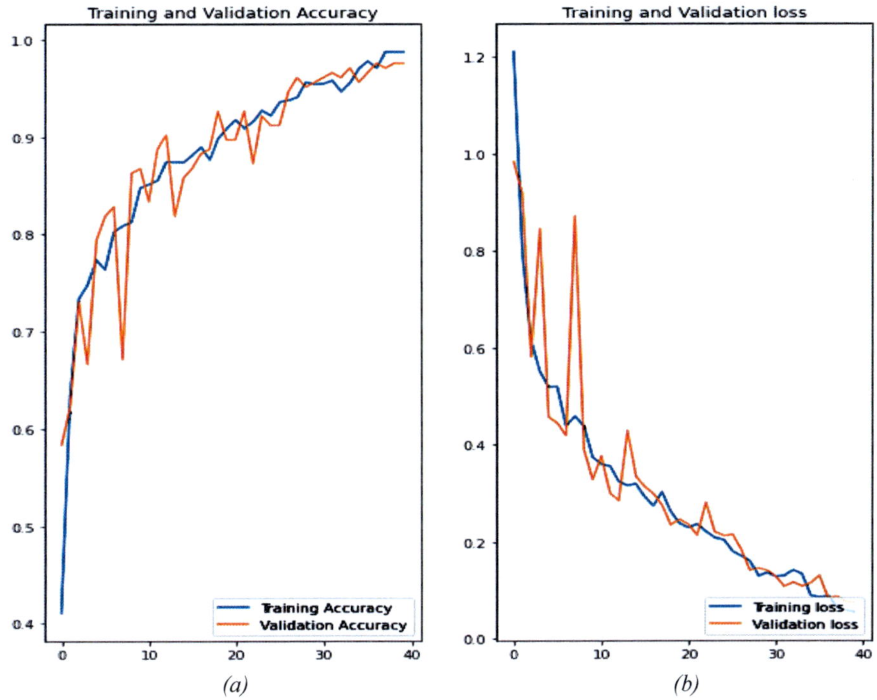

Fig. 4. (a) Training accuracy vs. validation accuracy. (b) Training loss vs. validation loss

9 images of early-stage disease out of 156 were misclassified as middle stage and healthy stage. It can be noticed further that 4 middle stage out of 142 and two end stage images out of 98 were not correctly classified as the true class.

True Class \ Predicted Class	Healthy Stage	Early Stage	Middle Stage	End Stage
Healthy Stage	140	1		
Early Stage	5	147	3	
Middle Stage		8	138	2
End Stage			1	96

Fig. 5. Confusion matrix for four classes

4 Conclusion

This study proposes a novel network for diagnosing plant diseases and estimating severity of wheat stem rust. It creates an end-to-end pipeline for diagnosing the severity of plant disease by automatically discovering discriminative characteristics for fine-grained categorization. The developed model outperforms, with a test set accuracy of 96.42%, proving that deep learning is a promising new technology for fully automatic classification of plant disease severity. Therefore, the presented framework is a viable candidate for use in a portable device that can diagnose crop diseases in real time. In the future, more images of various diseases at different severity levels can be collected to increase model accuracy. Additionally, after the model has been trained, it must be evaluated on images from a variety of sources in order to gain a better understanding of the model's actual utility. A precise estimation of disease severity could lead to the proper application of pesticides in the fields. Hyperspectral imaging combined with deep learning may be a promising method for early prediction, reducing the use of pesticides on crops significantly. The deep learning model can be used to forecast yields, make treatment recommendations, and so on. Moreover, the development of a mobile application for plant disease severity estimation could profit the farmers to overcome the technology barrier present and crop loss. The authors anticipate that the proposed framework would be enhanced to make a remarkable contribution to agricultural sciences.

References

1. Bock CH, Poole GH, Parker PE, Gottwald T (2010) R: Plant disease severity estimated visually, by digital photography and image analysis, and by hyperspectral imaging. Crit Rev Plant Sci 29(2):59–107
2. Mutka AM, Bart R (2015) S: Image-based phenotyping of plant disease symptoms. Front Plant Sci 5:734
3. Mohanty SP, Hughes DP, Salathé M (2016) Using deep learning for image-based plant disease detection. Front Plant Sci 7:1419
4. Yang X, Guo T (2017) Machine learning in plant disease research. European Journal of BioMedical Research. 3(1):6–9
5. Sapna N, Jain R (2020) Plant disease identification using deep learning: a review. Indian J Agric Sci 90(2):249–257
6. Schmidhuber J (2015) Deep learning in neural networks: an overview. Neural Netw 61:85–117
7. LeCun Y, Bengio Y, Hinton G (2015) Deep learning. Nature 521(7553):436
8. Kamilaris A, Boldú PFX (2018) Deep learning in agriculture: a survey. Comput Electron Agric 147:70–90
9. Too EC, Yujian L, Njuki S, Yingchun L (2019) A comparative study of fine-tuning deep learning models for plant disease identification. Comput Electron Agric 161:272–279
10. Wang G, Sun Y, Wang J (2017) Automatic image-based plant disease severity estimation using deep learning. Comput Intell Neurosci 2017. Article ID 2917536. https://doi.org/10.1155/2017/2917536
11. Pethybridge SJ, Nelson SC (2015) Leaf Doctor: A new portable application for quantifying plant disease severity. Plant Dis 99(10):1310–1316
12. Liang Q, Xiang S, Hu Y, Coppola G, Zhang D, Sun W (2019) PD2SE-Net: computer-assisted plant disease diagnosis and severity estimation network. Comput Electron Agric 157:518–529

13. Barbedo JGA (2018) Impact of dataset size and variety on the effectiveness of deep learning and transfer learning for plant disease classification. Comput Electron Agric 153:46–53
14. Szegedy C, Vanhoucke V, Ioffe S, Shlens J, Wojna Z (2016) Rethinking the inception architecture for computer vision. In: Proceedings of the IEEE conference on computer vision and pattern recognition, pp 2818–2826
15. LeCun Y, Bengio Y (1995) Convolutional networks for images, speech, and time series. In: The handbook of brain theory and neural networks, vol 10, p 3361
16. Lee SH, Chan CS, Wilkin P, Remagnino P (2015) Deep-plant: plant identification with convolutional neural networks. In: International conference on image processing, pp 452–456. IEEE
17. Nigam S, Jain R, Marwaha S, Arora A (2021) 12 wheat rust disease identification using deep learning. In: Internet of things and machine learning in agriculture, pp 239–250. De Gruyter

Check for updates

Credit Card Fraud Detection Using CNN

Yogamahalakshmi Murugan(✉), M. Vijayalakshmi, Lavanya Selvaraj, and Saranya Balaraman

Thiagarajar College of Engineering, Madurai, Tamilnadu, India
yoga@student.tce.edu

Abstract. In financial transactions, credit cards are increasingly popular, while fraud is growing as well. Traditional methods are focused on regulations based on experts that ignore a variety of conditions and an extreme discrepancy between positive and negative samples to detect fraud behaviors. A feature matrix, a neural network of convolutions, is used to characterize large quantities of transaction data. To identify a group of secret patterns, a network is used. Any specimen Real-world scenario experiments. The large transactions of a big commercial bank show their superior effectiveness over other cutting-edge approaches.

Keywords: Transaction data · Convolutional neural network · Trading entropy

1 Introduction

Credit card transaction datasets are infrequently accessible, highly distorted, and highly imbalanced. The most advantageous attribute selection for the models, worthy measure is the most important part of data mining to assess the performance of techniques. In credit card number of challenges are linked to credit card detection, that is, fraudulent activity profile which is complex. Fraudulent transactions appear to look like legalized transactions. The type of sampling method, variable selection, and detection techniques used all have a significant impact on performance [3].

We create a novel trading function called trading entropy that is focused on and customer's most recent consumption preferences. To suit a convolutional neural network (CNN) model to credit card fraud detection, we must first translate features into a feature matrix. Furthermore, highly imbalanced data is a problem in fraud detection. A common technique for adjusting the minority ratio is random under-sampling method for dominated classes.

Regrettably, it would necessarily ignore important facts. In this paper, we use a cost-based sampling approach to create synthetic fraudulent samples from real frauds. As a result, we get a similar amount of fraud and legal transactions for training purposes [13].

In a nutshell, the following are the key contributions to this paper:

1. We suggest a CNN-based method for mining credit card transaction latent fraud trends.
2. We turn each transaction's data into a feature matrix [4].

R. Misra et al. (Eds.): ICIoTCT 2021, LNNS 340, pp. 194–204, 2022.
https://doi.org/10.1007/978-3-030-94507-7_19

3. Allowing the CNN model to see the inherent flaws.
4. In periodicity, there are partnerships and experiences.
5. The extremely imbalanced sample sets are alleviated by integrating the cost-dependent sampling approach in characteristic space, resulting in superior fraud detection effectively.
6. To distinguish more complex financial instruments, a new trading function called trading entropy is proposed [2].

2 Related Work

"(John O. Awoyemi; Samuel A. Oluwadare 2017)" [11] The proposal looks at the performance of the ships, the closest neighbor and the logistically modified details on credit card fraud. ("Chee Bhagyani; Sasitha Premadasa et al. 2019") [7] In real-world transactions, they proposed four distinct methods of fraud. A selection of machine-learning models is used to deal with each fraud, and after evaluating, the best solution is selected. (Praveen Kumar Sadineni 2020") [15] To identify fraudulent transactions, they suggest using machine learning techniques such as Artificial Neural Networks (ANN), Decision Trees, Support Vector Machines (SVM), Logistic Regression, and Random Forest [1].

("L. Breiman, J. H. Friedman, R. A. et al. 2019") [6] They gave a quick overview of binary decision trees and demonstrated the findings in classifying Landsat-TM and AVIRIS digital images. ("Wen-Fang Yu; Na Wang 2019") [14] They propose a model of credit card fraud detection focused on the distances between the rare and irregular findings of credit card fraud, and use outer mining for the detection of credit card fraud. ("Ekrem Duman, Ilker Elikucuk 2013") [9] They suggested scattering search and genetic algorithms. Then they use a newly built metaheuristics algorithm called the migrating bird's optimization algorithm (MBO), which outperforms the previous method.

("Massimiliano Zanin, Miguel Romance, et al. 2018") [12] They proposed the first hybrid data mining/complex network classification algorithm capable of detecting illicit transactions in a real-world data set of card transactions. It is based on a recently proposed network reconstruction algorithm that allows representations of a single instance's divergence from a reference group to be created. ("Thulasyammal Ramiah Pillai, et al. 2018") [10] Using deep learning algorithms, they proposed a high-performance model for detecting credit card fraud. They discovered that both the logistic and hyperbolic tangent activation functions are effective in detecting credit card fraud.

("Devika S, Nisarga K, 2019") [8] They suggested an HMM-based credit card fraud identification system that looked at each card's fraudulent activity trends [15]. The majority of transactions are not verified by the investigators due to time and expense restraints. As a result, the transaction goes unnoticed before the client discovers the fraud report or until enough time has passed that the transaction is considered non-disputed honestly. ("P.K. Chan, et al. 2019) [5] They suggested general and demonstrably useful methods for integrating several studied fraud detectors under a "cost paradigm".

3 Proposed Method

Credit card firms square ready to accept illegitimate credit card purchases. Customers are not paying with products that they do not need. During this project, we tend to square

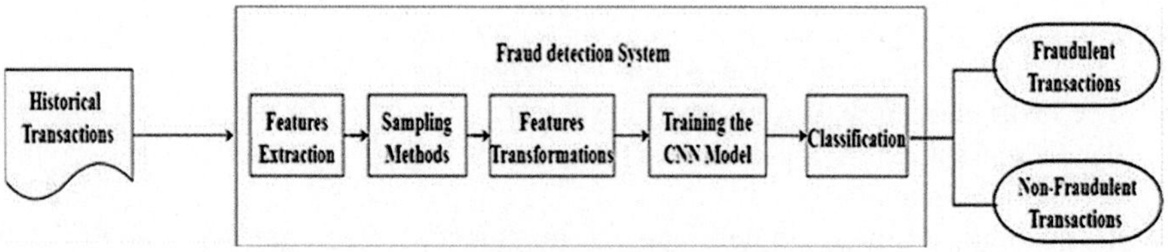

Fig. 1. Flow diagram

measure aiming to build a model exploitation CNN that predicts if the dealing is real or fraudulent. Figure 1 depicts the overview of the proposed system.

The phases in the methodology are as follows:

a) Data Collection
b) Pre-processing
c) Constructing CNN
d) The classification system

3.1 Data Collection

We use the credit card fraud detection dataset from Kaggle. It contains anonymized Credit card transactions tagged as fallacious or real. We will transfer it from there. Credit card transactions are included in the datasets. In September 2013, in European cardholders' data collection we have got 492 transactions that happened in two days. There were 284,807 deals of fraud. The amount of data being collected is enormous. In an imbalanced situation, the positive category (frauds) accounts for a significant portion of the total 0.0172% of all transactions.

3.2 Pre-processing (Balance Dataset)

Data pre-processing is a data mining technique that entails converting raw data into a format that can be understood. Real-world data is often incomplete, unreliable, deficient in some habits or patterns, and is likely to contain several errors. Pre-processing data is a tried-and-true way of addressing such problems. Pre-processing raw data prepares it for further processing. Database-driven systems, such as customer relationship management and rule-based applications, use data pre-processing (like neural networks). Data pre-processing is important in deep Learning processes because it encodes the dataset in a way that the algorithm can understand and parse.

3.3 Constructing CNN

Day by day, computer vision evolves at a breakneck rate. Deep learning is one of the reasons for this. The word convolutional neural network (abbreviated as CNN) comes to mind when we speak about computer vision since it is widely used. Face recognition, image classification, and other applications of CNN in computer vision are examples.

It works in the same way as a simple neural network. CNN, like neural networks, has learnable parameters. The layers in CNN is shown in Fig. 2.

Layers in CNN:

1. Input layer
2. Convo layer (Convo + ReLU)
3. Pooling layer
4. Fully connected (FC) layer
5. SoftMax/logistic layer
6. Output layer

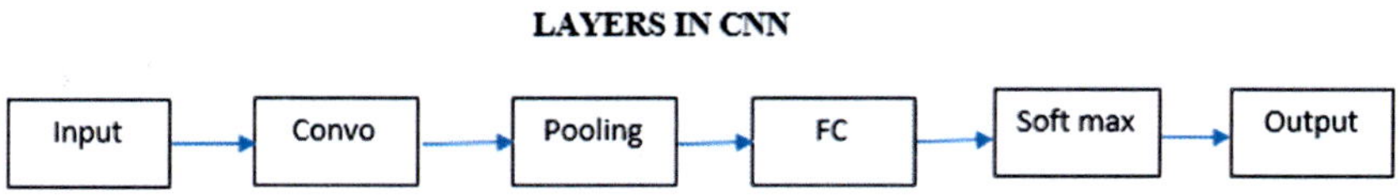

Fig. 2. Layers in CNN

4 Experiments and Result

4.1 Dataset

It does have numerical input variables that are squared. Calculate the outcome of a PCA transition. The options that haven't been reworked with PCA square measure 'Time' and 'Amount'. Each deal is separated by a second in the 'Time' feature, therefore the first deal in the dataset may be found in this feature. The 'Number' function is for dealings quantity, and it will be used for examples. Cost-sensitive learning is a form of learning that is based on the cost. The answer variable 'Class' is the function, and it takes price one only in the case of fraud and zeros otherwise (Fig. 3).

4.2 Importing Tensorflow and Keras

We square measure progressing to use TensorFlow to create the model. You will install TensorFlow by running this command. If your machine encompasses a GPU you'll use the second command.

TensorFlow is employed to create the neural network. We have even foreign all the layers needed to create the model from Keras. NumPy is employed to perform basic array operations. pandas for loading and manipulating the info. The plot from Matplotlib is employed to ascertain the results. train_test_split is employed to separate the info into coaching and testing datasets. StandardScaler is employed to scale the values within the information.

	Time	V1	V2	V3	V4	V5	V6	V7	V8	V9	...
0	0.0	-1.359807	-0.072781	2.536347	1.378155	-0.338321	0.462388	0.239599	0.098698	0.363787	...
1	0.0	1.191857	0.266151	0.166480	0.448154	0.060018	-0.082361	-0.078803	0.085102	-0.255425	...
2	1.0	-1.358354	-1.340163	1.773209	0.379780	-0.503198	1.800499	0.791461	0.247676	-1.514654	...
3	1.0	-0.966272	-0.185226	1.792993	-0.863291	-0.010309	1.247203	0.237609	0.377436	-1.387024	...
4	2.0	-1.158233	0.877737	1.548718	0.403034	-0.407193	0.095921	0.592941	-0.270533	0.817739	...

...	V21	V22	V23	V24	V25	V26	V27	V28	Amount	Class
...	-0.018307	0.277838	-0.110474	0.066928	0.128539	-0.189115	0.133558	-0.021053	149.62	0
...	-0.225775	-0.638672	0.101288	-0.339846	0.167170	0.125895	-0.008983	0.014724	2.69	0
...	0.247998	0.771679	0.909412	-0.689281	-0.327642	-0.139097	-0.055353	-0.059752	378.66	0
...	-0.108300	0.005274	-0.190321	-1.175575	0.647376	-0.221929	0.062723	0.061458	123.50	0
...	-0.009431	0.798278	-0.137458	0.141267	-0.206010	0.502292	0.219422	0.215153	69.99	0

Fig. 3. Sample dataset

4.3 Balanced Dataset

Here we will produce a variable non_fraud which can contain the information of all the real transactions. The transactions with [‘Class’] == 0. fraud can contain the information of all the dishonest transactions i.e., the transactions with [‘Class’] == 1. The form attribute tells the USA that non_fraud has 284315 rows and thirty-one columns and fraud has 492 rows and thirty-one columns.

To balance the information, we will choose 492 transactions at random from non_fraud. Now you can see that non_fraud has 492 rows. Now we will produce the new balanced dataset by appending non_fraud to fraud. As for ignoring index = True, the ensuing axis is going to be tagged zero, 1, …, n − 1.

We will separate the featured house and therefore the category. X can contain the featured house and y can contain the category label. Now we will split the information into coaching and testing set with the assistance of train_test_split(). test size = zero.2 can keep two 0% information for testing and eightieth information is going to be used for coaching the model. The random state controls the shuffling applied to the information before applying the split. Stratify = y implies that the information is split in a very stratified fashion, victimization y because of the category labels.

We can see that there are square measure 787 samples for coaching and 197 samples for testing. StandardScaler() standardizes the options by removing the mean and scaling to unit variance. We will work pulse counter solely to the coaching dataset however we transform each coaching also because of the testing dataset.

Our information is two-dimensional however neural networks settle for three-dimensional information. thus, we've to reshape the information using reshape() (Fig. 4).

Time	V1	V2	V3	V4	V5	V6	V7	V8	V9	...
406.0	-2.312227	1.951992	-1.609851	3.997906	-0.522188	-1.426545	-2.537387	1.391657	-2.770089	...
472.0	-3.043541	-3.157307	1.088463	2.288644	1.359805	-1.064823	0.325574	-0.067794	-0.270953	...
4462.0	-2.303350	1.759247	-0.359745	2.330243	-0.821628	-0.075788	0.562320	-0.399147	-0.238253	...
6986.0	-4.397974	1.358367	-2.592844	2.679787	-1.128131	-1.706536	-3.496197	-0.248778	-0.247768	...
7519.0	1.234235	3.019740	-4.304597	4.732795	3.624201	-1.357746	1.713445	-0.496358	-1.282858	...

...	V21	V22	V23	V24	V25	V26	V27	V28	Amount	Class
...	0.517232	-0.035049	-0.465211	0.320198	0.044519	0.177840	0.261145	-0.143276	0.00	1
...	0.661696	0.435477	1.375966	-0.293803	0.279798	-0.145362	-0.252773	0.035764	529.00	1
...	-0.294166	-0.932391	0.172726	-0.087330	-0.156114	-0.542628	0.039566	-0.153029	239.93	1
...	0.573574	0.176968	-0.436207	-0.053502	0.252405	-0.657488	-0.827136	0.849573	59.00	1
...	-0.379068	-0.704181	-0.656805	-1.632653	1.488901	0.566797	-0.010016	0.146793	1.00	1

Fig. 4. Balanced dataset

4.4 Constructing CNN

A Sequential model is acceptable for a clear stack of layers wherever every layer has precisely one input tensor and one output tensor. Conv1D() may be a 1D Convolution Layer. This layer is incredibly effective for explanation options from a fixed-length phase of the dataset. Wherever it is not vital, wherever the feature is found within the phase, within the 1st Conv1D() layer we tend to learn a complete of thirty-two filters with a size of the convolutional window as a pair of input shape specifies the form of the input. It is a necessary parameter for the primary layer in any neural network. We are going to be victimization ReLu activation perform. The corrected linear activation perform or ReLu for brief may be a piecewise linear perform that maybe 1 if the input is positive; otherwise, it will be 0.

Batch Normalization() permits every layer of a network to be told by itself a touch bit a lot of severally of different layers. to extend the steadiness of a neural network, batch normalization normalizes the output of a previous activation layer by subtracting the batch mean and dividing by the batch variance. It applies a metamorphosis that maintains the mean output getting ready to zero and also the output variance getting ready to one.

Dropout() is employed to necessarily set the outgoing edges of hidden units to zero at every update of the coaching part. The worth passed in dropout specifies the likelihood at which those outputs of the layer is born out.

Flatten() is employed to convert the info into a 1-dimensional array for inputting it to a future layer. The standard strongly connected neural network layer is Dense(). The output layer is additionally a dense layer with one vegetative cell as a result of we tend to at predicting one price as this is often a binary classification downside. Sigmoid perform

U.S.A.edlis employed} as a result of it exists between (0 to 1) and this facilitates us to predict a binary input.

Now we are going to compile and work on the model. We use a victimization Adam optimizer with a 0.00001 learning rate. we are going to use twenty epochs to coach the model. associate degree epoch is associate degree iteration over the complete information provided. At the top of each epoch, validation data is the details used to judge the failure and some model metrics. As metrics = ['accuracy'] the model is going to be evaluated to support the accuracy (Figs. 5 and 6).

```
Layer (type)                 Output Shape              Param #
=================================================================
conv1d_2 (Conv1D)            (None, 29, 32)            96
_________________________________________________________________
batch_normalization_2 (Batch (None, 29, 32)            128
_________________________________________________________________
dropout_3 (Dropout)          (None, 29, 32)            0
_________________________________________________________________
conv1d_3 (Conv1D)            (None, 28, 64)            4160
_________________________________________________________________
batch_normalization_3 (Batch (None, 28, 64)            256
_________________________________________________________________
dropout_4 (Dropout)          (None, 28, 64)            0
_________________________________________________________________
flatten_1 (Flatten)          (None, 1792)              0
_________________________________________________________________
dense_2 (Dense)              (None, 64)                114752
_________________________________________________________________
dropout_5 (Dropout)          (None, 64)                0
_________________________________________________________________
dense_3 (Dense)              (None, 1)                 65
=================================================================
Total params: 119,457
Trainable params: 119,265
Non-trainable params: 192
_________________________________________________________________
```

Fig. 5. Layer conversion model

4.5 Plotting Accuracy and Loss Graph

We can see that the coaching accuracy is above the validation accuracy. Therefore, we will say that the model is overfitting. We add a Max Pool layer and increase the number of epochs to enhance our accuracy (Fig. 7).

4.6 Adding Max-Pool

Now we are going to once more visualize the results. It can be seen that we get a stronger result after we re-train our model with several changes (Figs. 8 and 9).

4.7 Confusion Matrix

We have predicted the confusion matrix using classes, for fraud as 1 and for non-fraud as 0 (Fig. 10).

```
Epoch 1/10
25/25 [==============================] - 7s 60ms/step - loss: 0.6612 - accuracy: 0.6874 - val_loss: 0.6106 - val_accuracy: 0.82
74
Epoch 2/10
25/25 [==============================] - ETA: 0s - loss: 0.4559 - accuracy: 0.82 - 1s 25ms/step - loss: 0.4518 - accuracy: 0.82
72 - val_loss: 0.5591 - val_accuracy: 0.8223
Epoch 3/10
25/25 [==============================] - 1s 20ms/step - loss: 0.3833 - accuracy: 0.8615 - val_loss: 0.5199 - val_accuracy: 0.87
31
Epoch 4/10
25/25 [==============================] - 1s 23ms/step - loss: 0.3311 - accuracy: 0.8755 - val_loss: 0.4838 - val_accuracy: 0.90
36
Epoch 5/10
25/25 [==============================] - 1s 21ms/step - loss: 0.3612 - accuracy: 0.8767 - val_loss: 0.4453 - val_accuracy: 0.90
86
Epoch 6/10
25/25 [==============================] - 1s 21ms/step - loss: 0.2753 - accuracy: 0.9009 - val_loss: 0.4103 - val_accuracy: 0.90
86
Epoch 7/10
25/25 [==============================] - 1s 22ms/step - loss: 0.2662 - accuracy: 0.9022 - val_loss: 0.3788 - val_accuracy: 0.91
88
Epoch 8/10
25/25 [==============================] - 1s 21ms/step - loss: 0.2450 - accuracy: 0.9123 - val_loss: 0.3414 - val_accuracy: 0.91
37
Epoch 9/10
25/25 [==============================] - 1s 21ms/step - loss: 0.2402 - accuracy: 0.9250 - val_loss: 0.3205 - val_accuracy: 0.92
39
Epoch 10/10
25/25 [==============================] - 1s 24ms/step - loss: 0.2296 - accuracy: 0.9263 - val_loss: 0.2874 - val_accuracy: 0.91
88
```

Fig. 6. Accuracy and loss rate

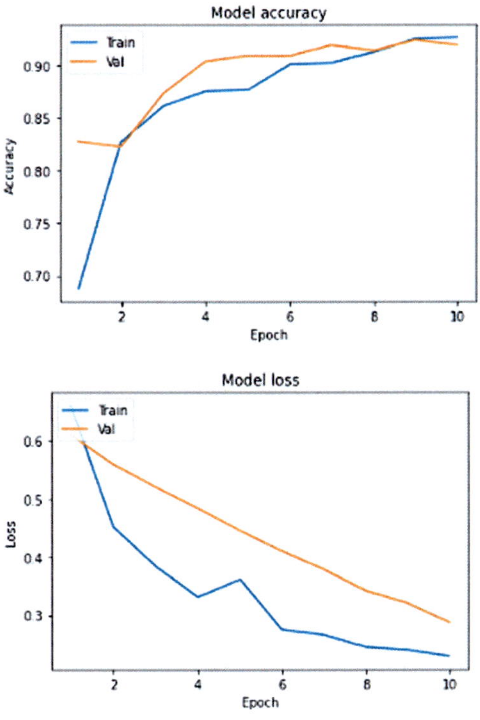

Fig. 7. Accuracy and loss graph

```
Epoch 1/10
25/25 [==============================] - 3s 34ms/step - loss: 0.4986 - accuracy: 0.8539 - val_loss: 0.4184 - val_accuracy: 0.87
31
Epoch 2/10
25/25 [==============================] - 0s 17ms/step - loss: 0.2884 - accuracy: 0.9060 - val_loss: 0.3170 - val_accuracy: 0.91
37
Epoch 3/10
25/25 [==============================] - 0s 16ms/step - loss: 0.2348 - accuracy: 0.9187 - val_loss: 0.2295 - val_accuracy: 0.91
88
Epoch 4/10
25/25 [==============================] - 0s 17ms/step - loss: 0.2083 - accuracy: 0.9263 - val_loss: 0.2067 - val_accuracy: 0.90
36
Epoch 5/10
25/25 [==============================] - 0s 17ms/step - loss: 0.2067 - accuracy: 0.9327 - val_loss: 0.2067 - val_accuracy: 0.91
88
Epoch 6/10
25/25 [==============================] - 0s 18ms/step - loss: 0.2125 - accuracy: 0.9327 - val_loss: 0.2098 - val_accuracy: 0.91
88
Epoch 7/10
25/25 [==============================] - 0s 16ms/step - loss: 0.1711 - accuracy: 0.9365 - val_loss: 0.1995 - val_accuracy: 0.91
88
Epoch 8/10
25/25 [==============================] - 0s 18ms/step - loss: 0.1777 - accuracy: 0.9352 - val_loss: 0.1862 - val_accuracy: 0.91
37
Epoch 9/10
25/25 [==============================] - 1s 21ms/step - loss: 0.1701 - accuracy: 0.9390 - val_loss: 0.2494 - val_accuracy: 0.90
86
Epoch 10/10
25/25 [==============================] - 0s 18ms/step - loss: 0.1655 - accuracy: 0.9517 - val_loss: 0.2080 - val_accuracy: 0.91
88
```

Fig. 8. Accuracy and loss rate after adding max pool

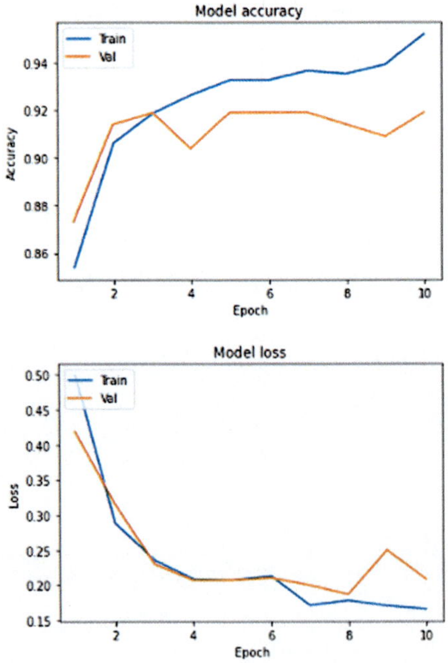

Fig. 9. Accuracy and loss graph for after adding max pool

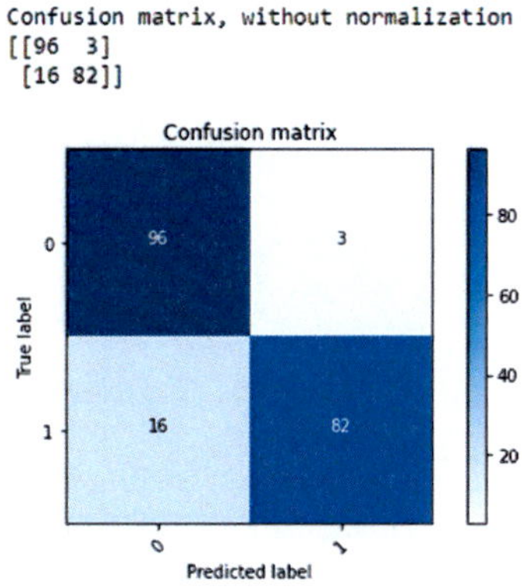

Fig. 10. Confusion matrix

4.8 Performance Metrics

We have used the confusion matrix to predicted the performance metrics like precision, recall, f1-score and support value (Fig. 11).

	precision	recall	f1-score	support
0	0.86	0.97	0.91	99
1	0.96	0.84	0.90	98
accuracy			0.90	197
macro avg	0.91	0.90	0.90	197
weighted avg	0.91	0.90	0.90	197

Fig. 11. Performance metrics

5 Conclusion

For this proposed method we have got more accuracy and less error rate in the existing method. We have used credit card transactions in the deep learning approach method using a Convolutional neural network.

In this paper, we tend to introduce a CNN-based methodology of credit card fraud detection and the trading entropy is planned to model a lot of advanced overwhelming behaviors. Besides, we tend to recombine the trading options to feature matrices and use them in an exceedingly convolutional neural network. The important transaction's experimental findings knowledge from a bank reveals that our expected methodology outperforms state-of-the-art alternative ways.

References

1. Srivastava A, Kundu A (2018) credit card fraud detection using hidden Markov model. IEEE Trans Dependable Secure Comput 5(1):37–48

2. Shen A, Tong R et al (2017) Application of classification models on credit card fraud detection. In: International conference on service systems and service management
3. Portia A, Raj BE (2011) Analysis on credit card fraud detection methods. In: International conference on computer, communication and electrical technology (ICCCET)
4. Ambeth Kumar VD, Kumar A et al (2019) credit card fraud detection using data analytic techniques. Adv Math Sci J 9(3):1185–1196
5. Chan P, Fan W, Prodromidis AL et al (2019) Distributed data mining in credit card fraud detection
6. Breiman L, Friedman JH et al (1987) Classification and regression trees, Wadsworth, Belmont, CA
7. Bhagyani C, Kuruwitaarachchi N et al (2019) Real-time credit card fraud detection using machine learning. In: 9th the international conference on cloud computing, data science & engineering (confluence)
8. Chandini SB, Devika SP et al (2019) A research on credit card fraudulent detection system. Int J Recent Technol Eng (IJRTE) 8:5029–5032
9. Duman E, Elikucuk I (2013) Solving credit card fraud detection problem by the new meta-heuristics migrating birds optimization. In: Rojas I, Joya G, Cabestany J (eds) IWANN 2013, vol 7903. LNCS. Springer, Heidelberg, pp 62–71. https://doi.org/10.1007/978-3-642-38682-4_8
10. Hashem IAT et al (2018) Credit card fraud detection using deep learning technique. In: Fourth international conference on advances in computing, communication & automation (ICACCA)
11. Awoyemi JO, Oluwadare SA et al (2017) Credit card fraud detection using machine learning techniques: comparative analysis. In: International Conference on computing networking and informatics (ICCNI)
12. Zanin M, Romance M et al (2018) Credit card fraud detection through parenclitic network analysis
13. Hacid M-S, Zeineddine H et al (2019) an experimental study with imbalanced classification approaches for credit card fraud detection. IEEE Access 7:93010–93022
14. Wang N, Yu W-F (2009) Research on credit card fraud detection model based on distance sum. In: International joint conference
15. Sadineni PK (2020) Detection of fraudulent transactions in credit card using machine learning algorithms. In: Fourth international conference on I-SMAC (IoT in social, mobile, analytics, and cloud) (I-SMAC)

Familial Analysis of Malicious Android Apps Controlling IOT Devices

Subhadhriti Maikap, Pushkar Kishore(✉), Swadhin Kumar Barisal, and Durga Prasad Mohapatra

NIT Rourkela, Rourkela, Odisha, India
518cs1002@nitrkl.ac.in

Abstract. The Android Operating System is the usual and well-known medium for accessing and controlling IoT devices since 2012. As android gets popular, malware has become a part of it. The obfuscated malware and its detection method have improved significantly, but traditional detectors are unsuccessful in detecting them. This paper suggests a framework for identifying malicious Android apps with features extracted using static and dynamic analysis. Experiments are performed on over 8000 applications (benign and malware) which can communicate with IoT devices. Furthermore, we experiment and assess our model on the off-the-shelf dataset, namely Drebin. The experimental study reveals that our proposed model can reach 96.7% detection accuracy, which outperforms traditional techniques. Apart from that, the familial accuracy is 82.62%, which suggests that classification is appropriately done by the detector and better than recent works. Furthermore, the classification and detection results justify the choice of static and dynamic analysis (hybrid analysis) features over the single analysis.

Keywords: Android · IoT · Hybrid analysis · Familial analysis

1 Introduction

Starting from 2012, we use the well-known operating system, android, for accessing and controlling IoT devices. In the year 2021, the estimated market share of android was 84%[1]. With the popularity of android, a centralized application marketplace like Google Play Store has grown massively, making smartphone applications (Apps) a prevalent medium for managing personalized computing services like electronic mail, banking, gaming, media consumption, IoT devices, etc. However, countless mobile apps' massive popularity and availability have caused the android operating system vulnerable to various malware intrusions. Android is open-source, which means it is not limited to commercial Apps, which led to the growth of a vast number of third-party applications. These third-party apps make up a significant element of the Android experience and also makes the Android platform an obvious target for malware intrusions.

[1] https://www.idc.com/promo/smartphone-market-share.

R. Misra et al. (Eds.): ICIoTCT 2021, LNNS 340, pp. 205–214, 2022.
https://doi.org/10.1007/978-3-030-94507-7_20

Forbes organization concluded a study which attested that malicious samples related to the android operating system covered 97% of the total mobile malware[2]. According to a current report published in 2020, Kaspersky mobile security engine and malware detection technologies identified 5,683,694 malicious installation packages on Android[3]. Google Play Store has its security mechanism called Google Play Protect, which scans and verifies more than 50 billion android packages every day. The organization named McAfee reported that google play protect failed in detecting a few malware found in 2017[4]. Consequently, we need an effective and continuous research for discovering and blocking those zero-day attack malware.

Analysis tools accomplish the static and dynamic analysis of app one by one, and the outcomes of analysis determine the dataset's features. We can validate the app at the design phase of project development for bugs and security issues [8]. System calls are used for defining the behavior of the samples [12]. Then, we combine them to create a set of features that can train the malware detector. Finally, the hybrid analysis is accomplished on the feature-set mentioned above, and a framework is proposed to provide higher detection and familial accuracy.

Paper Organizations: The paper is split among sections and organized as follows: Sect. 2 briefly explains Android Application Package, Sect. 3 explains the related work, Sect. 4 presents our proposed model's methodology, Sect. 5 manifests the experimental outcomes, Sect. 6 examines the comparison with related work and Sect. 7 substantiates the conclusions and future work.

2 Android Application Package

Android app is developed and stored in the format android application package (APK) for distribution. It is an archived file containing all the components needed to run the application. In addition, this file holds all the program's codes (classes.dex), assets, resources, manifest files, and certificates. For example, Fig. 1 represents the APK file structure discussed above.

1. *DEX file:* After compilation, the android program generates .dex (Dalvik Executable) files; then, they are zipped into a .apk file. These .dex files are similar to java class files; they mainly store the compiled java classes and run under the Dalvik Virtual Machine (DVM).
2. *Manifest file:* Android Manifest file (AndroidManifest.xml) is an XML formatted object required by every Android application. When an app is launched on the device, the android operating system first searches for the manifest. In addition, the file provides firsthand information about the security settings and characteristics of the application.

---

[2] https://www.forbes.com/sites/gordonkelly/2014/03/24/report-97-of-mobile-malware.

[3] https://www.kaspersky.com/resource-center/threats/mobile.

[4] https://www.mcafee.com/blogs/other-blogs/mcafee-labs/android-malware-grabos-exposed-millions-to-pay-per-install-scam-on-google-play.

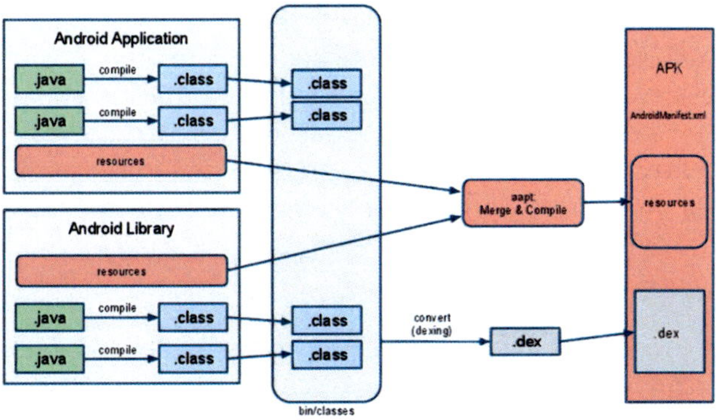

Fig. 1. APK file structure

3 Related Work

This section includes state-of-the-art works by several researchers and enterprises.

3.1 Static Analysis

The .apk files are reverse-engineered, then the .dex files and AndroidManifest.xml are analyzed. The analysis extracts a log file (most commonly a .json file) that contains the package's static features such as permissions, intents, services, activities, receivers, etc. Peiravian et al. [1] applied a machine learning (ML) based approach on static features. They extracted two types of features, namely user permissions and API function call. Wang et al. [2] designed 'Droid-DeepLearner' a malware discovery model based on user permissions. Finally, RiskRanker [3] recognized Android apps having various security risks through a static approach.

3.2 Dynamic Analysis

In dynamic analysis, the .apk package is installed on an actual device or an emulator. Dash et al. [4] designed a dynamic analysis based model for android malware detection. They had extracted four behavioral features, namely Network access, File access, Binder Methods, and Executed files with Support Vector Machine (SVM) for classification. DroidRanger [5] and AppsPlayground [6] proposed to investigate android apps and detected possible malicious activities dynamically.

3.3 Hybrid Analysis

We combine the static and dynamic analysis to create a hybrid one. Alzaylaee et al. [7] designed a hybrid model for android malware discovery and trained using deep learning techniques. They had built an automated platform based on Android Virtual Device (AVD) to execute android apps and obtain their features. They had considered both static, dynamic, stateless, and stateful features as well. They had taken a total of 420 features. They applied the information gain metric to rank the features and performed feature selection. Yuan et al. [9] proposed 'DroidDetector', a hybrid approach for android malware detection.

3.4 Familial Analysis

Classifying malicious samples into their respective families (Familial Analysis) is essential in malware investigation and threat evaluation. It aids analysts in recognizing malware that is arising from a related source and holds similar malicious intents. RevealDroid [10] used sensitive API flow tracking and information flow analysis based on two ML classifiers (C4.5 and 1NN). Dendroid [11] proposed a model based on control flow structure and used text mining to categorize malicious samples automatically. DroidSIFT [13] built dependency graphs for API calls requested by the app. The graph's similarity was compared with other dependency graphs and encoded into feature vectors to recognize malware classes.

4 Proposed Methodology

This section describes our proposed hybrid (static + dynamic) model to detect and classify android malware using Random Forest (RF) and Deep Learning (DL) techniques. First, we present the architecture of the proposed malware detector in Fig. 2. It incorporates three steps: Dataset Collection, Static & Dynamic Analysis, and Feature Extraction & Feature Processing. Finally, we manifest the algorithm in Algorithm 1. The description of the above three steps is discussed below.

4.1 Dataset Collection

To evaluate our model performance, we replaced few samples of the dataset[5] with IoT device targeting Apps and finally selected 8287 Android applications. Out of which, 4304 are benign, and 3983 are malicious samples. The details of the malware families are discussed in Table 1.

4.2 Static and Dynamic Analysis

In AndroPyTool [14], we have FlowDroid for static analysis and DroidBox for dynamic analysis. For static analysis, the classes.dex and the AndroidManifest.xml files are analyzed to log static features. For dynamic analysis, the android application package is installed on the emulator (AVD), and dynamic behaviors are observed while running the application. Both static and dynamic analysis generates JSON files containing static and dynamic behavior, respectively.

[5] https://www.kaggle.com/goorax/static-analysis-of-android-malware-of-2017.

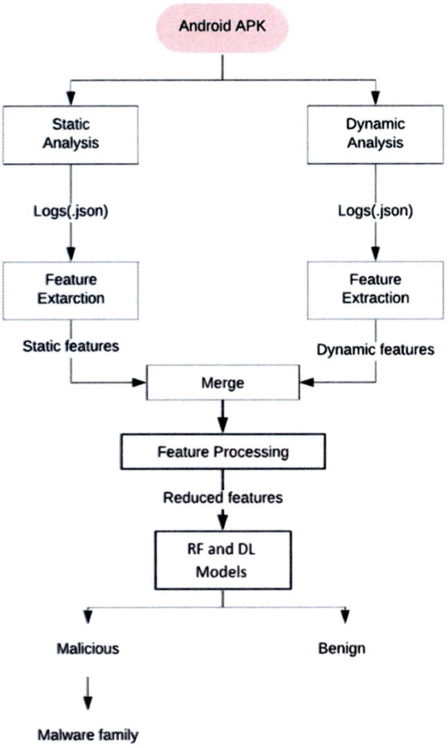

Fig. 2. Proposed architecture of our approach

4.3 Feature Extraction and Processing

From the JSON files obtained using the static and dynamic analysis, we perform feature extraction. We have extracted a total of 3075 features with 2988 static and

Table 1. Considered malware

Serial number	Family	Samples
1	Virus	2556
2	Trojan	168
3	Spyware	31
4	Risktool	579
5	Scareware	37
6	Ransomware	46
7	Downloader	44
8	SMSmalware	42
9	Adware	390
10	Dropper	75
11	Rootkit	15
Total number of malware		3983

Table 2. Feature set

Type	Feature category	Number of features
Static	User permissions	1053
Static	Activities	952
Static	Intents	783
Static	Services	75
Static	Receivers	46
Static	Providers	46
Static	Usefeatures	33
Dynamic	API calls during runtime	87
Total features		3075

87 dynamic features. The sample list of features is given in the following Table 2. All the features may not contribute equally to enhance the performance of our model; thus, irrelevant features must be ignored. To perform feature selection, we use Pearson correlation. After removing the irrelevant features, we obtain a feature set of 1024 features, out of which 966 are static features, and 58 are dynamic features.

5 Experimental Setup and Results

We present our model in detail in this section and emphasize our proposed approach's experimental results. RF and DL techniques are used for detection and familial analysis.

5.1 Experimental Setup

To execute android applications and extract their features, an automated platform is required. Since we aim to analyze android apps statically and dynamically, we utilize the AndroPyTool [13] hybrid analysis framework. The CPU version is Intel i5-8265U, RAM is 8 GB, and OS is Windows 10.

5.2 Evaluation Using RF

In our first experiment, we evaluate our model with a random forest. RF algorithm is a supervised characterization model. RF makes multiple decision trees on various subsets of the given dataset and merges them to get a more accurate and stable prediction. The more trees in the forest, the higher the accuracy and also prevents the overfitting problem. Their cumulative effort reduces the individual error of each tree. By evaluating the RF algorithm, we achieve a detection accuracy of 96.7%. Detailed results of the RF model are given in Table 3.

Algorithm 1: Hybrid analysis based android malware detection and familial analysis

Input: A set of android applications (APKs), APKs = $\{\mathrm{APK_b}, \mathrm{APK_m}\}$, where $\mathrm{APK_b}$ represents benign android application and $\mathrm{APK_m}$ represents malicious android application

Output: A final label informing whether app is malicious or benign along with the family

Function `AndMalDetectAndClassify`($APKs$):
- **for** *each apk in the APKs* **do**
  - *FlowDroid* ← Perform static analysis of the benign and malicious android packages and obtain analysis report;
- **for** *each apk in the APKs* **do**
  - *DroidBox* ← Perform dynamic analysis of the benign and malicious android packages and obtain analysis report;
- **for** *each analysis report from benign samples* **do**
  - $Feature_{\mathrm{s}}$ ← Perform feature extraction from static analysis report;
- **for** *each analysis report from malicious samples* **do**
  - $Feature_{\mathrm{d}}$ ← Perform feature extraction from dynamic analysis report;
- **for** *each features in* $Feature_b$ **do**
  - $Feature_{\mathrm{s}}$ ← Select most relevant features from static features;
- **for** *each features in* $Feature_m$ **do**
  - $Feature_{\mathrm{d}}$ ← Select most relevant features from dynamic features;
- Merge Static features and Dynamic features;
- $Features \leftarrow Feature_{\mathrm{s}} + Feature_{\mathrm{d}}$;
- *AndMalDet* ← Apply the *Features* to train Deep Learning based classification Models;
- **for** *each APK in the dataset* **do**
  - Apply *AndMalDet* classifier to predict the final label and malicious family;
- **return**;

end

5.3 Evaluation Using DL

For ascertaining the target of the app, we choose customized architecture, namely LeNet-5. We investigate the model on the accuracy parameter as it estimates the values of true positives, true negatives, false positives, and false negatives. The test sample's determination result is summarized in Table 3.

5.4 Familial Analysis of All Samples

The classification accuracy achieved by DL is 82.62% which is better than RF. Adware is correctly classified but sometimes misclassified as Smsmalware due to similarity in the way of attacking hosts. Spyware and Smsmalware are intended

Table 3. Performance evaluation of malware detection

Performance parameters	RF	DL
Accuracy	0.967	0.91
Precision	0.942	0.91
Recall	0.95	0.91
TNR (%)	96.49	88.6
FPR (%)	3.5	11.4
FNR (%)	4.6	7.6
F1-score	0.945	0.91
Detection duration (s)	1.5	12
Training duration (s)	120	825

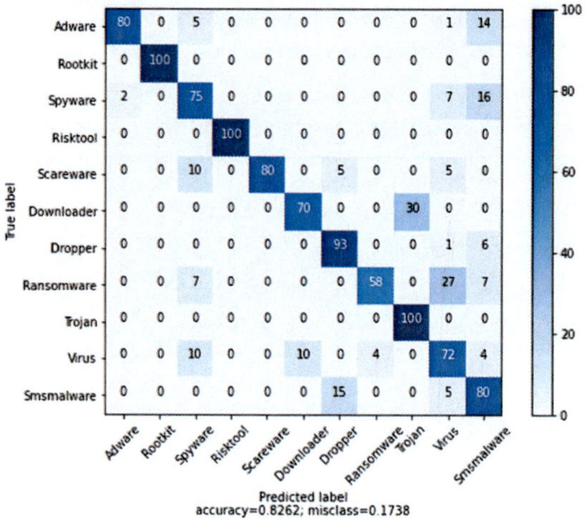

Fig. 3. Confusion matrix of classification of malware using deep learning

to steal private data, which led to the sharing of few features. Due to the above reason, the spyware gets mislabeled. A few Scareware samples get misclassified as spyware, while some downloader samples a trojan. Downloader is a mixture of various types of malware, thus gets mislabeled as trojan or virus sometimes. Classification of Ransomware is not highly effective and gets mislabeled as a virus due to its higher similarity with it. The trojan is correctly classified, while some virus instances get mislabeled as spyware and downloader. A few Smsmalware gets mislabeled as dropper since it can be downloaded by dropper.

6 Comparison with Related Work

In this section, we compare the performance of the proposed model with several works shown in Table 4. Our approach focuses on malware classification,

Table 4. Comparison of performance of proposed model with existing approaches

Method	Acc. (%)	Prec. (%)	Rec. (%)	F1-score (%)	DT(s)	TT(s)
SVM [1]	95.75	91.7	95.7	93.66	0.001	15
J48 [1]	94.46	90.6	92.8	91.69	0.005	134
Bagging [1]	96.39	94.9	94.1	94.5	NA	NA
DDL-DBN [2]	NA	93.09	94.5	93.71	0.094	179
DDL-SVC [2]	NA	97.59	87.73	88.74	NA	NA
DD-Static [9]	89.03	90.39	89.04	89.76	NA	NA
DD-Dynamic [9]	71.25	72.59	71.25	71.92	NA	NA
DD-Static+Dynamic [9]	96.62	95.6	97.6	96.58	NA	NA
DD-DL [15]	93.68	93.69	93.36	93.68	NA	NA
Our approach	96.7	94.2	95.0	94.5	1.5	120

and we consider more malware families than DroiScribe [4], and DroidDeep [2]. Overall our model has higher detection accuracy and needs the least training time. DL-Droid [7] had higher detection accuracy, but they accomplished only the detection part of malware and left behind the classification. We have higher classification accuracy using a less time-consuming model, RF, than a more time-consuming model, DL. Thus, some classes can be detected quickly using RF. However, familial analysis is better using DL instead of RF. Upon comparison, we perceive that the proposed model has the highest accuracy and is preferable for malware discovery and classification. Our model's computational complexity is almost the same as Bagging [4] and DDL [2].

7 Conclusions and Future Work

In this paper, we present a hybrid analysis model for android malware detection and familial analysis. We perform our experiment on over 8000 applications (benign and malware) which can communicate with IoT devices. The experimental study reveal that our proposed model can achieve 96.7% detection accuracy with Random Forest outperforming traditional techniques. The familial accuracy is 82.62%, which suggests that the DL module properly does classification. Thus RF is effective for detection and DL for classification.

In the future, we plan to further reduce the false positives and negatives by analyzing the reason behind the misclassification of samples. We will also assess the performance using other DL techniques.

References

1. Peiravian N, Zhu X (2013) Machine learning for android malware detection using permission and API calls. In: 25th international conference on tools with artificial intelligence, pp 300–305

2. Wang Z, Cai J, Cheng S, Li W (2016) Droiddeeplearner: identifying android malware using deep learning. In: 37th Sarnoff symposium, pp 160–165
3. Grace M, Zhou Y, Zhang Q, Zou S, Jiang X (2012) Riskranker: scalable and accurate zero-day android malware detection. In: 10th international conference on mobile systems, applications and services, pp 281–294
4. Dash SK, Suarez-Tangil S, Khan S, et al (2016) Droidscribe: classifying android malware based on runtime behavior. IEEE security and privacy workshops (SPW), pp 252–261
5. Shaerpour K, Dehghantanha A, Mahmod R (2013) Trends in android malware detection. J Digit Forensics Secur Law 8:21
6. Rastogi V, Chen Y, Enck W (2013) Appsplayground: automatic security analysis of smartphone applications. In: Proceedings of the third ACM conference on Data and application security and privacy, pp 209–220
7. Alzaylaee MK, Yerima SY, Sezer S (2020) DL-Droid: deeplearning based android malware detection using real devices. Comput. Secur 89:101663
8. Barisal SK, Behera SS, Godboley S, Mohapatra DP (2019) Validating object-oriented software at design phase by achieving MC/DC. Int J Syst Assur Eng Manag 10(4):811–823
9. Yuan Z, Lu Y, Xue Y (2016) Droiddetector: android malware characterization and detection using deep learning.'Tsinghua Sci Technol 21(1):114–123
10. Garcia J, Hammad M, Malek S (2018) Lightweight,obfuscation-resilient detection and family identification of android mal-ware. In: IEEE/ACM 40th international conference on software engineering (ICSE), pp 497–507
11. Suarez-Tangil G, Tapiador JE, Peris-Lopez P, Blasco J (2014) Dendroid: a text mining approach to analyzing and classifying code structures in android malware families. Exp. Syst Appl 41(4):1104–1117
12. Kishore P, Barisal SK, Vaish S (2019) Nitrsct: a software security tool for collection and analysis of kernel calls. In: IEEE region 10 conference (TENCON), pp 510–515
13. Zhang M, Duan Y, Yin H, Zhao Z (2014) Semantics aware android malware classification using weighted contextual API dependency graphs. In: CCS 2014: proceedings of the 2014 ACM SIGSAC conference on computer and communications security, pp 1105–1116. Association for Computing Machinery, New York
14. Martin Garcia A, Lara-Cabrera R, Camacho D (2018) A new tool for static and dynamic android malware analysis. In: Conference on data science and knowledge engineering for sensing decision support (FLINS 2018), pp 509–516
15. Hou S, Saas A, Chen L, Ye Y (2016) Deep4maldroid: a deep learning framework for android malware detection based on Linux kernel system call graphs. In: IEEE/WIC/ACM international conference on web intelligence workshops (WIW), pp 104–111

SERI: SEcure Routing in IoT

Varnika Gaur, Rahul Johari, Parth Khandelwal(✉), and Apala Pramanik

SWINGER (Security, Wireless, IoT Network Group of Engineering and Research) Lab, USICT, GGSIP University, Sector-16C, Dwarka, Delhi, India

Abstract. In the world of data communication and information security, safe and secure transmission of the message over the insecure network is a challenge. Various researchers around the world have devised various techniques, algorithms and methodologies to ensure secure and reliable sharing of the message between source and destination in the network. In the proposed work, a message is encrypted using DES (Data Encryption Standard) Algorithm and is routed using MQTT (Message Queuing Telemetry Transport) in IoT (Internet of Things). The entire approach has been conceptualized and programmed using JCA (Java Cryptography Architecture) API's of Java Programming language and is executed using Paho Client-IoT-ECLIPSE IDE.

Keywords: IoT · Cryptography · Security · MQTT · DES · Routing

1 Introduction

As well known,the last decade has seen the emergence of incredible and amazing technologies such as Internet of Things(IoT), BlockChain Technology, Artificial Intelligence including Machine Learning and Deep Learning, MICEF [Mist, Internet of Things (IoT), Cloud, Edge and Fog Computing], Quantum computing, Big Data tools and technologies et al. which are helping the programmers and researchers to design, develop and deploy world class innovative products for the benefit of mankind.

With broadband internet being available widely and economical price, the cost of connecting devices has gone down significantly accompanied by increased accessibility of smart phones has created the perfect environment for the inevitable growth of Internet of things. Internet of things refers to any device (that has on/off switch) and can be connected to the internet. IoT is a network of interconnected devices and people which interact with each other in order to share information and data. IoT devices are embedded with sensors, software and other requisite technology to facilitate the transfer of data over the internet. These can include any device from household items to sophisticated industrial tools. IoT devices collect data regarding there surrounding environment. The rapid evolution of IoT has led to internet penetrating every aspect of everyday

R. Misra et al. (Eds.): ICIoTCT 2021, LNNS 340, pp. 215–224, 2022.
https://doi.org/10.1007/978-3-030-94507-7_21

life through embedded systems. Experts estimate that there would be 22 billion IoT devices by the year 2025. The concept of IoT was first given by Kevin Ashton in 1999, he defined IoT uniquely identifiable interoperable interconnected devices with radio frequency identification technology. In the last decade or so the applications the far-reaching applications have been realised in almost every sector from retail, logistics to pharmaceuticals. The success of IoT has a direct co-relation to the standardisation, which provides interoperability, compatibility and reliability for operations on global scale. One of the most critical requirements of IoT is that all things on the network are interconnected and IoT architecture should guarantee seamless bridging of the gap between physical and virtual worlds. The design of architecture for IoT takes into consideration the extensibility, scalability, interoperability of multiple heterogeneous devices along with considering their business models. Today IoT is an important component of SHIP (Smart-Sparse, Hybrid, IoT and Partitioned) Network (Fig. 1) as millions of smart and intelligent electronic devices, fitted with sensors are now available for use, for the customers.

The suitable architecture for IoT is Service oriented Architecture (SoA), which ensures interoperability among heterogeneous devices in multiple ways. SoA consists of four layers:

1. Sensing layer is integrated with available hardware objects to sense the status of things.
2. Network layer provides the infrastructure to support over wireless or wired connections among things.
3. Service layer is to create and manage services required by users or applications.
4. Interface layer consists of the interaction methods with users or applications.

2 Applications of IoT

- **Wearable devices**: the devices such as virtual glass, GPS tracking belts, calorie counting, step monitor are just few of the examples of the wearables. These devices are equipped with the necessary hardware, software and sensors to collect and store information about the user.
- **Healthcare**: These devices allow doctors to monitor patients in real time outside of the hospital facilities. These devices provide regular updates on the patient's health and come equipped with alarm systems in case patient is in distress. They help in improving care of patients and help prevent fatal events in case of high- risk patients.
- **Traffic monitoring**: The apps such as Google maps utilises our mobile phones as sensors to pin our location on map and suggests a suitable route based on feedback regarding traffic.
- **Power Supply**: Installation of intelligent sensors at strategic points that go from power plants to distribution plants help better monitoring and control of the electrical network. The installation of bidirectional communication between the supplier and final consumers can help detect faults and repair

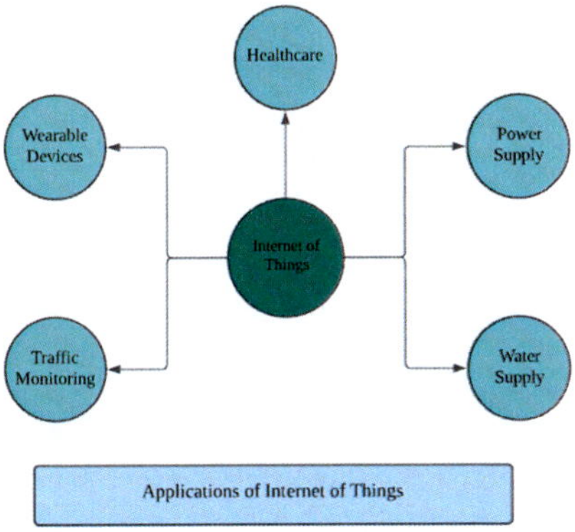

Fig. 1. Various applications of IoT

thereof. These devices can also help in tracking consumption of the households and advise consumers on how to manage their electricity expenditure.

- **Water Supply**: A sensor incorporated in water metres along with the necessary software and internet connection, collects data regarding consumption of water by individual households and generates reports regarding consumer behaviour and helps in detecting leaks in the supply. It also generates reports regarding average water consumption and helps final consumers track their consumption reports.

3 Objective

The Primary objective of undertaking current research work is to showcase message encryption using DES (Data Encryption Standard) Algorithm which is routed using MQTT (Message Queuing Telemetry Transport) in IoT (Internet of Things). The entire approach has been conceptualized and programmed using JCA (Java Cryptography Architecture) API's of Java Programming language and is executed using Paho Client-IoT-ECLIPSE IDE.

4 Literature Survey

Lounis et al. [1] discussed a simulation tool CupCarbon, which is based on primarily two constituents. First is the multi agent simulation environment which, with the help of a user-friendly interface of OpenStreetMap framework permits of replication an setting analogous to the actual world where each element is implemented autonomously and simultaneously. Second is the wireless network (WSN) simulator which allows to design a network of agents such as sensors, mobiles etc. and then simulate the events associated to these agents. Sensors can be directly positioned on the map and thus various parameters can be monitored for changes over a period of time. The paper presents a case study where the energy diagram associated to the sensors in the simulation of "destructive" insect movements is obtained. The network comprises of eight sensors that have been allotted a single script and simulation parameters are: the simulation time and the simulation step.

Bounceur et al. [2] describes a platform called CupCarbon-Lab built on the basis of the CupCarbon simulator where network of objects, linked via internet, can be effortlessly simulated, in analogy to the real world. This simulation tool not only allows to check the viability of various algorithms but also directly instil the code into the hardware and reconstruct it as per requirement. The author shows an example of a SenScript that is used to program the nodes of the simulator and the various 2D and 3D illustrations of a simulated network of sensor nodes.

Bounceur et al. [3] presents the newly proposed architecture of the already existing CupCarbon platform which is primarily used for the design creation, visualization and simulation of giant wireless sensor networks based on IOT. The author explains the architecture i.e. the four key blocks of the simulation tool, with diagrams, that are: Radio Channel Block, Inference Block, 2D/3D Environment Block and Implementation Block and further elucidates the various modules of the projected version. Thus the paper concludes with a new form of the existing CupCarbon simulation tool which allows the computer-generated nodes to be substituted by the genuine ones. It also advocates that using this version, any physical IOT network can be implemented without having to code any actual sensor node exclusively and thus various parameters of the real network can be monitored precisely.

López-Pavón et al. [4] evaluated the CupCarbon software for the simulation of wireless sensor networks (WSNs) and discussed the primary features of the simulator including the user interface. The Dijkstra algorithm which computes the cheapest path between two points has been implemented and simulated with a variation in order to examine the performance of the simulator. An additional parameter i.e. battery level of the nodes has been added as a modification to the traditional algorithm and the results have been shown.

Ojie et al. [5] discussed the importance Internet of Things (IoT) in making lives of human beings much more convenient. Smart cities and hospitals are some examples of how IoT has transformed the lives of people and how such networks require minimum human involvement to run. The authors have highlighted the need for simulation and testing in order to check the performance and efficiency before actually implementing a complex system and have therefore evaluated the various tools available for the purpose of simulation. Each of these tools have been discussed and tabulated according to their features and the functions they can perform. An analysis has been given based on the comparison between the features and a chart has been presented with the advantages and disadvantages of using each tool.

Bounceur et al. [6] presented a new geometric calculation based D-LPCN algorithm that can detect gaps or uncovered areas in wireless sensor networks (WSNs) installed in areas which cannot be accessed by humans. Voids occur when the network is deployed randomly and these voids might lead to serious issues like cyber security attacks if not detected. The authors have used a new approach to describe a polygon as interior or exterior, that is, the minimum x-coordinate based method, where polygon global minimum (PGM) denotes the vertex of polygon with smallest x-coordinate.

D'Angelo et al. [7] discussed the concerns and problems, both qualitative and quantitative, regarding simulation and testing of large networks of sensors and devices before actually installing them in the real world. In order to deploy large systems of Internet of Things (IoT), such as smart cities that have a dense network of nodes, the authors have proposed new simulation techniques which would not only improve scalability but also allow real time simulation of such large networks in performance evaluation. Further, to achieve this, the authors have proposed a "parallel and distributive simulation (PADS)" approach along with multi-level simulation.

Capponi et al. [8] discussed the parameters to be considered while designing a Mobile Crowdsensing (MCS) simulator. Moreover, two key performance indicators (KPI) that must be evaluated have been discussed which are, data generation and cost evaluation. Thus, on the basis of the design principals and suggested KPIs, a new simulator prototype has been developed in order to deliver a simulation environment for MCS systems. Based on the energy consumed by the devices and the data generated, preliminary results have been discussed.

Chernyshev et al. [9] presented an overall comparative study on the simulation tools that are currently used by researchers for the examination of IoT prototypes. A review of the trends in IoT research activities have been presented for a period of five years, that is, from 2012 to 2017 considering the publications under four main publishers. Research goals that would ensure successful implementation of IoT in future has been discussed. For deployment of large IoT environments, simulation and testing is an extremely important step. Thus, numerous types of existing simulators currently used by researchers have been discussed and compared on the basis of three broad categories. First category is of Full Stack Simulators such as e DPWSim, and iFogSim. Second is Big Data Processing Simulators such as IOTSim and SimIoT. Third category is of Network Simulators such as CupCarbon, Cooja, OMNeT++, NS-3, and QualNet. Further, the authors have highlighted the importance of open IoT test beds and discussed three test bed, FIT IoT-LAB, SmartSantander and Japan-Wide Orchestrated Smart/Sensor Environment (JOSE).

Gheisari et al. [10] discussed one of the primary issues of IoT devices which is privacy preservation of data while performing analysis or transferring from unauthorized users. Privacy can be divided into two categories that are, data privacy and context privacy and the attacks that can take place might be internal or external. Thus to protect data, the authors have discussed the need for privacy preserving data publishing (PPDP) and proposed a novel method, Modular Arithmetic Algorithm for Privacy Preservation (MAPP) which is based on number theory and modular arithmetic.

Lavric et al. [11] discussed the LoRaWAN communication protocol used in IoT and analysed its performance and sustainability. The network has been simulated on the CupCarbon as a solution to the scalability issue of LoRanetwork. This simulator which allows users to simulated discrete events and wireless sensor networks. SenScript is used to program at the node level. In this paper, 500 LoRa nodes were connected and simulated. The results show a variation in frequency of the LoRa modulated signal.

Johari et al. [12] discussed the importance of routing in IoT and the requirement of studying and developing protocols for routing in IoT. Some of the advantages of IoT, as discussed in the paper, include, enhanced customer satisfaction, digital optimization and waste minimization. Tracking and reducing energy consumption, healthcare industry, education purpose and government projects are some of the domains where IoT can prove to be useful.

Pramanik et al. [13] proposed a smart helmet for the visually challenged persons in the society. This helmet alerts the people around the handicapped person whenever he is about to take a turn or stops moving. Moreover it vibrates to alert the user when any vehicle or person comes close to that person.

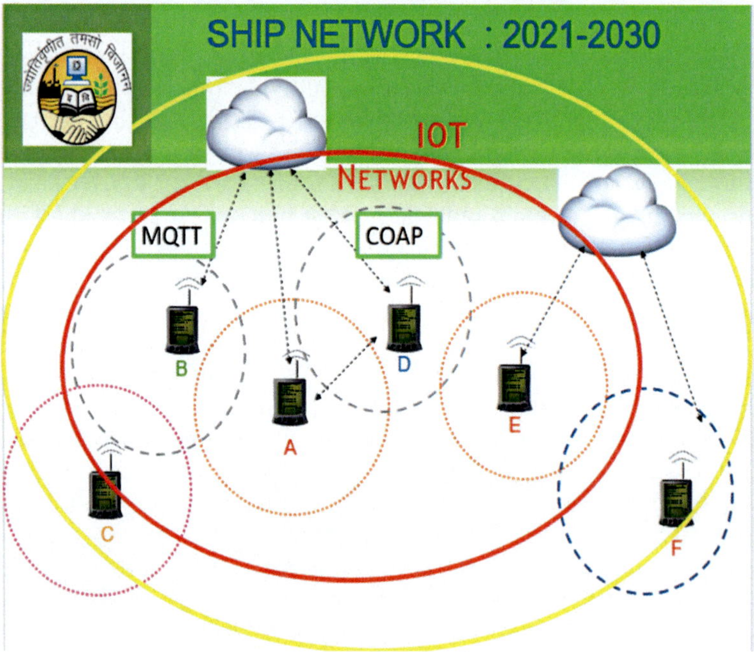

Fig. 2. Message routing using MQTT and COAP protocol

5 Proposed Approach

In IoT, to deal with routing of the messages, the publisher-subscriber model is widely used. In the proposed work the plaintext is accepted from the user which is Maximum temperature of the Day in the month of May 2021. The user just enters the numeric value of the atmospheric temperature at the run-time and it is concatenated with the constant string, for example: Numeric Value + "Degree Celsius on 21 May 2021 in Delhi". This plaintext is then encrypted using DES (Data Encryption Standard) Algorithm by the publisher. The Ciphertext is transmitted to the broker by the publisher. The Ciphertext is then routed from the IoT Broker to the subscriber using the MQTT (Message Queuing Telemetry Transport) protocol. The Subscriber then decrypts the ciphertext to obtain back the original plaintext which is '33 Degree Celsius on 21 May 2021 in Delhi'.

5.1 Experimental Setup

See Table 1.

Table 1. Hardware and software used in simulation

S. no	Hardware and Software requirements	Description
1	Operating system	WINDOWS
2	MQTT version number	1.5.4
3	MQTT port number	8883
4	JDK version	JDK 1.8.0
5	CPU processor	Intel i5

6 Result

The program was written in Java Programming language to simulate the scenario where in the Publisher-Broker-Subscriber can seamlessly communicate with each using the MQTT Protocol on the secured port number 8883. The Experimental set is shown in Fig. 3 and the snapshot of the MQTT Server, Encryption-Decryption of the Plaintext using DES Algorithm and the entries in log trace file of the MQTT server are shown in Fig. 3, 4 and 5 respectively.

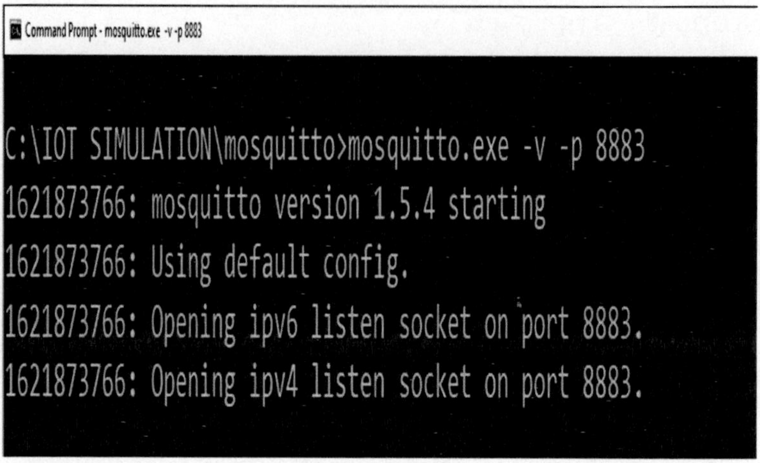

Fig. 3. Snapshot of the start of the IoT MQTT server on port number 8883

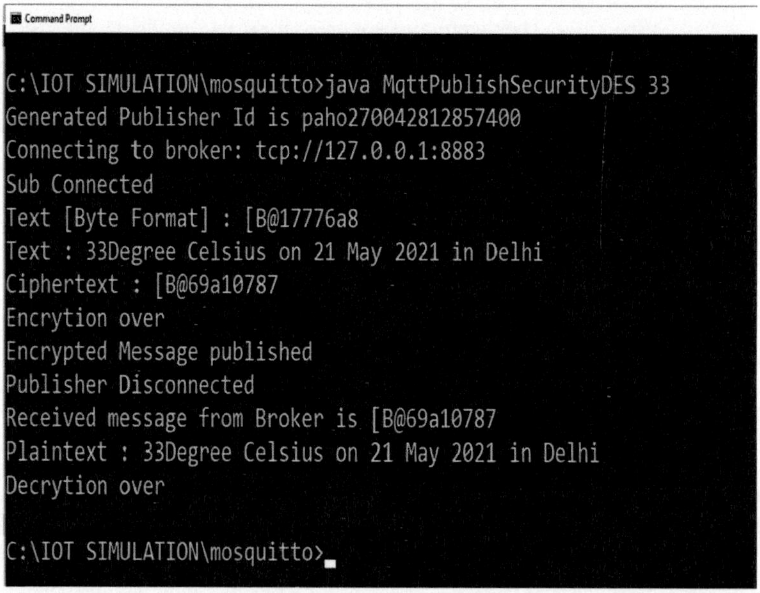

Fig. 4. Snapshot of the encryption of the plaintext using DES algorithm exploiting publisher-subscriber model of IoT

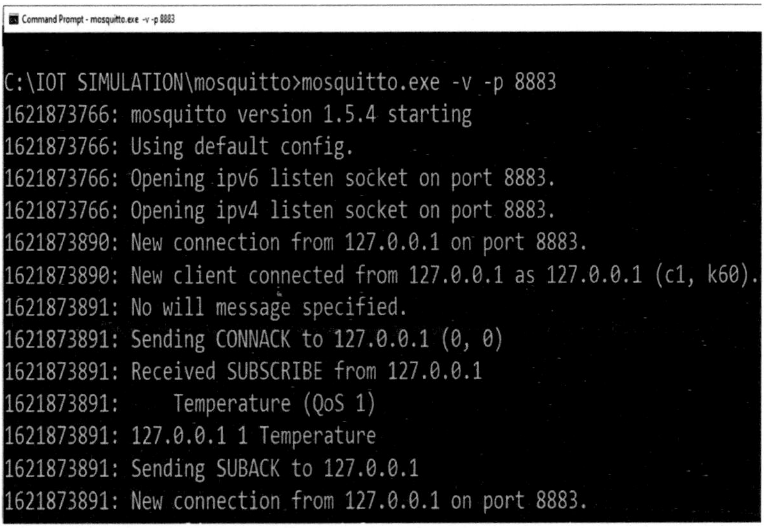

Fig. 5. Snapshot of the log file traces at the MQTT server

7 Conclusion

In the work presented in the current research paper, a plaintext message was successfully encrypted using DES (Data Encryption Standard) Algorithm, routed using MQTT (Message Queuing Telemetry Transport) exploiting the Publisher-Subscriber model in IoT. The entire approach was conceptualized and programmed using JCA (Java Cryptography Architecture) API's of Java Programming language and was executed using Paho Client-IoT-ECLIPSE IDE. In the future it is planned to enhance the security in IoT by using more Symmetric and Asymmetric Cryptographic Algorithms like AES (Advanced Encryption Standard) and RSA respectively.

References

1. Lounis M, Mehdi K, Bounceur A (March 2014) A CupCarbon tool for simulating destructive insect movements. In: 1st IEEE international conference on information and communication technologies for disaster management (ICT-DM 2014), Algiers, Algeria
2. Bounceur A, et al (January 2018) CupCarbon-Lab: an IoT emulator. In: 2018 15th IEEE annual consumer communications & networking conference (CCNC). IEEE, pp 1–2
3. Bounceur A, et al (January 2018) CupCarbon: a new platform for the design, simulation and 2D/3D visualization of radio propagation and interferences in IoT networks. In: 2018 15th IEEE annual consumer communications & networking conference (CCNC). IEEE, pp 1–4
4. López-Pavón C, Sendra S, Valenzuela-Valdés JF (2018) Evaluation of CupCarbon network simulator for wireless sensor networks. Netw Prot Alg 10(2):1–27

5. Ojie E, Pereira E (October 2017) Simulation tools in internet of things: a review. In: Proceedings of the 1st international conference on internet of things and machine learning, pp 1–7
6. Bounceur A, et al (June 2018) Detecting gaps and voids in WSNs and IoT networks: the minimum x-coordinate based method. In: Proceedings of the 2nd international conference on future networks and distributed systems, pp 1–7
7. D'Angelo G, Ferretti S, Ghini V (July 2016) Simulation of the internet of things. In: 2016 international conference on high performance computing & simulation (HPCS). IEEE, pp 1–8
8. Capponi A, Fiandrino C, Franck C, Sorger U, Kliazovich D, Bouvry P (December 2016) Assessing performance of internet of things-based mobile crowdsensing systems for sensing as a service applications in smart cities. In: 2016 IEEE international conference on cloud computing technology and science (CloudCom). IEEE, pp 456–459
9. Chernyshev M, Baig Z, Bello O, Zeadally S (2017) Internet of things (IoT): research, simulators, and testbeds. IEEE Internet Things J 5(3):1637–1647
10. Gheisari M, Wang G, Bhuiyan MZA, Zhang W (December 2017) MAPP: a modular arithmetic algorithm for privacy preserving in IoT. In: 2017 IEEE international symposium on parallel and distributed processing with applications and 2017 IEEE international conference on ubiquitous computing and communications (ISPA/IUCC). IEEE, pp 897–903
11. Lavric A, Petrariu AI (May 2018) LoRaWAN communication protocol: the new era of IoT. In: 2018 international conference on development and application systems (DAS). IEEE, pp 74–77
12. Johari R, Adhikari S (February 2020) Routing in IoT network using CupCarbon simulator. In: 2020 7th international conference on signal processing and integrated networks (SPIN). IEEE, pp 301–306
13. Pramanik A, Johari R, Gaurav NK, Chaudhary S, Tripathi R (2021) ASTITVA: assistive special tools and technologies for inclusion of visually challenged. In: 2021 international conference on computing, communication, and intelligent systems (ICCCIS), pp 1060–1065. https://doi.org/10.1109/ICCCIS51004.2021.9397168

A Review Paper on Machine Learning Based Trojan Detection in the IoT Chips

T. Lavanya(✉) and K. Rajalakshmi

Department of Electronics and Communication Engineering, PSG College of Technology, Tamilnadu Coimbatore, India
lavanya.takur@gmail.com, krl.ece@psgtech.ac.in

Abstract. Recently, Internet of Things (IoTs) are in wide use and the IoTs are referred as growing network of devices which establish an internet connectivity and communication among the devices. The IoT uses billions of data as it connects billions of devices to the internet. The IoT has different layers such as sensing layer, physical layer and application layer, each layer undergoes various method of security. However, the basic fundamental of IoTs are hardware, thus there is a more concern to secure the hardware from the adversary attack as the Integrated Circuits (ICs) manufacturing flow are globalized. The adversary attacks may cause a malicious functions like change in the functionality of the circuit, leakage of data, reduce reliability, modification of parameters etc. hence, there is a need for securing the hardware this is known as Hardware Security. In recent times, many researchers as come up with various methodologies to secure the Hardware by detecting the presence of Trojan in the Hardware. One such methodology is the Machine Learning based Hardware Trojan detection, where various classifiers are trained to detect the small Trojans present in the Hardware circuit. Thus, in this article we are going to highlight the important aspects of Machine Learning Classifier over the Hardware Security problems. Here we are going to compare various Machine Learning Classifiers used so far in the Hardware Trojan Detection problem.

1 Introduction

In semiconductor technology, the advancement of IC design is growing enormously because of globalization. The design and the manufacturing process of ICs are vulnerable to attacks because of the third-party participants. The vulnerability in the device lead to the malicious attack on the device which alters the functionality of the device, leakage of data, etc. these threats are causing a serious concern in various applications such as household appliances, military systems, financial infrastructures, and etc. [1, 2]. The vulnerability in these processes led to the emergence of hardware security to secure the devices against the external threats as that of software security. The third-party vendor can introduce any trojans at any stage of ICs. Thus, these trojans affect the hardware device as per the requirement of adversary. The IC fabrication process and the threats involve three major steps as shown in Fig. 1 [3]: i) design-which includes IP, models, tools. The tools used in these designs are said to be from trusted vendors whereas IP,

R. Misra et al. (Eds.): ICIoTCT 2021, LNNS 340, pp. 225–238, 2022.
https://doi.org/10.1007/978-3-030-94507-7_22

models that is used by the designers may be from untrusted vendors. ii) fabrication- it includes mask generation, lithography and packaging, this fabrication process may also be considered as untrusted because there may be a chance of addition of unwanted circuit. iii) manufacturing test- to verify the manufactured IC, this can be trusted if its verification is done under the trusted unit. The threats during these processes is difficult to avoid as it consumes more time and expensive because of increase in manufacturing units of ICs and requirement of ICs globally.

Hardware Trojans are defined as an external circuit implemented as a hardware modification in ASICs, digital-signal processors, microprocessors, microcontrollers, etc. [1, 3] The trojan circuit consist of a payload logic and trigger logic in it as shown in Fig. 2 [3]. The trigger logic get activates when the adversary passes the signal externally this activates the payload logic, the activated payload logic performs an unwanted function in actual design thus leads to malfunctioning of the circuit. In order to avoid these unwanted actions, the hardware-based security techniques are eventually used to modify the hardware to prevent the [3] IP blocks, attacks and secret keys.

The attacks considered are of different types i.e., attack may be during fabrication or before fabrication. Detection of such modifications in the circuit is highly difficult because of various reasons [4]. Firstly, the usage of huge number of hard IP blocks, soft firms used in SoCs and the complexity of IP blocks have made the detection of minute modifications in the SoCs difficult. Secondly, the reverse engineering methods and physical inspection method are difficult as the ICs are sized in a nanometers scale, thus these methods are said to be costly and time consuming. Destructive reverse engineering is not that effective because it doesn't ensure remaining ICs are trojan free as the insertion of trojans is selectively in any of the chip. Thirdly, the trojan gets activated only for specified nets or signals as designed by the adversary to perform the malfunctioning in the circuit thus, it is difficult to detect the possible nets. Fourthly, the design for test can only detect the faults present in the netlist of the design i.e., it can detect stuck-at-faults but it does not ensure the design is trojan free as the nets used are to determine the functionality of the design. Thus, the trojan nets are difficult to determine [4, 5].

As the technology are increasing the [6] physical feature sizes of a design are decreasing as the lithography process, masking and environmental variations are improving in recent times, these increasing the impact on the integrity of circuit parameters. Thus, by knowing the parameters of the circuit by simple analysis the trojan detection are ineffective.

2 Taxonomy of Hardware Trojan

The taxonomy of trojan circuit [4, 7] is represented in various forms as it is evolving because of new attacks and new trojan types are discovered. The Fig. 3 shows the general taxonomy of hardware trojans which is based on five attributes. These attributes are given as follows:

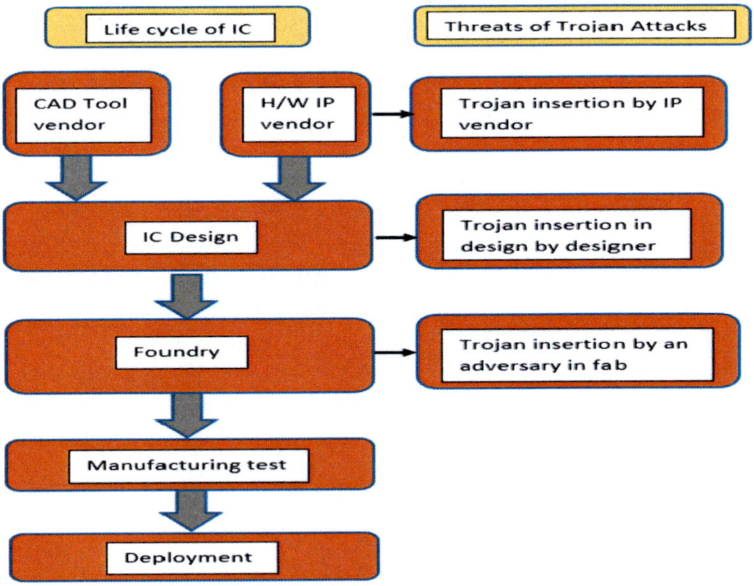

Fig. 1. Hardware Trojan attacks at different stages of IC process [3]

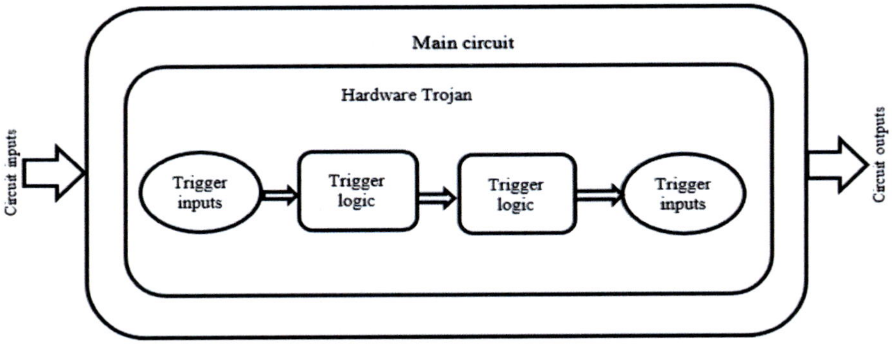

Fig. 2. Basic structure of hardware Trojan [3]

A. Insertion Phase

The trojans are classified based on the phase of insertion [8]. As the hardware trojan insertion may takes place in any development stage of the chip. Thus, it is necessary to know the stages of insertion of hardware trojan.

- **Specification phase:** in the specification phase, there may be possibilities of changes in the specific parameters like delay, power, function of a design, area etc. these parametric changes cause a damage to a targeted circuit which degrades the performance of the circuit.

- **Design phase:** in the design phase, the designer uses the available Intellectual Property (IP) blocks and other logic cells, gates, flipflops, etc. which are provided by the third-party vendor in designing the circuit. In this phase the mapping of functional, logical, timing constraints onto the required technology.
- **Fabrication phase:** in the fabrication phase, as the fabrication of chips undergo in an unknown foundry there is a possibility of attack by an adversary who can insert the Trojan through his own masks or through using his own wafers this may cause a serious effect on the chip. Here, there is a possibility of changing the chemical composition of the layers. This increase the power consumption of the device which in turn lead to the aging of the chip.
- **Assembly phase:** in this phase there is an assembling of 'n' number of components on a printed-circuit-board (PCB). The adversary on other hand may add up the additional components to the PCB in the beginning these components are inactive in nature thus even after testing it is unable to detect but as the time goes these components are activated and it affect the functionality of the original device.

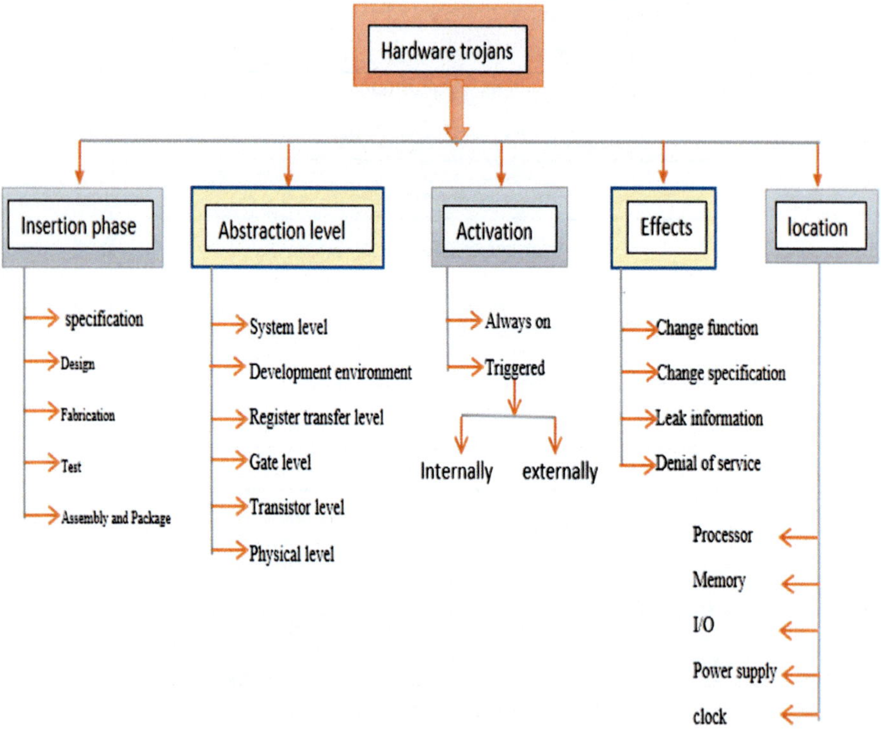

Fig. 3. Taxonomy of hardware Trojan [28]

- **Testing phase:** in this phase, after the assembling of the PCB the circuit is set to be tested before it is deployed to the market. The testing method are used to test the circuit even then it is difficult to identify the Trojan presence as the Trojan are inactive thus it is difficult to determine by testing methodology.

B. Level of description

This describes the level of presence of Trojans present in the circuit in the various abstraction level such as, system level, development environment, RTL level, gate level, transistor level, layout level. These levels of abstraction are discussed below as the functionality of the design depend on these levels [8].

- **System level:** during the design of a circuit various modules are used such as, interconnection modules, communication protocol blocks, hardware logic blocks etc. and these modules get triggered by the trigger signal sent by an adversary to activate the Trojan and perform a possible attack in the targeted circuit.
- **Development environment:** this indicates the nature of the environment used for designing the circuit such as simulation, synthesis, verification and validation tools. These tools are provided by various vendors and thus, there may be insertion of Trojan in CAD tools and scripts though software [3]. Hence it is difficult to identify the presence of Trojan.
- **Register transfer level:** the RTL level involves the functional modules such as registers, signals, and Boolean functions. The attackers make use of these modules thus it is prone to attack by an adversary thus the Trojan can be inserted in these modules.
- **Gate level:** in this level, the logic gates are used in a design and these gates are interconnected to perform a particular function of the design. The Trojan inserted by an adversary affect the size and location of the design thus there is a malfunction in the design. The Trojan inserted in these gate level may be of functional Trojan and they may be sequential or combination type of Trojan.
- **Transistor level:** the chip designer control over the circuit characteristics and to build a logic gates transistors are used. Thus, the attacker can alter the functionality of the circuit characteristics by addition or deletion of the transistor. To modify the parameter of the circuit the size of the transistor may be changed or the transistor is made to be always in active state. This causes a huge delay in the critical path of the circuit.
- **Layout level:** the layout level also called as physical level as it describes the dimension and location of each circuit parameters present in the design. In this level the attacker attacks by modifying the size of the wire, increase or decrease the distance between the components, alignment of various layers, overlapping of layers etc. thus even this level is prone to Trojan attack.

C. Activation mechanism

The activation mechanism [9] is the device how and when the Trojan gets activated based on the intention of the adversary. There are two different methods involved in activating the Trojan. One such is by the triggered signal i.e., once the triggered signal is sent the Trojan gets activated until that it will be in dormant state. Other is when the target circuit is said to be operational, i.e., as soon as the target circuit is powered ON the Trojan circuit will also said to be activated and it will be in always ON state. Thus, it seems to be more dangerous than the triggered ones as it leads to aging of the device. The triggered signal is of two variants one is internally triggered and the other is externally triggered based on these signals the Trojan gets activated and these Trojans in a circuit are in dormant state until it is triggered.

- **Internally triggered mechanism:** the internally triggered mechanism involves triggering the Trojan based on the occurrence of any specified event internally. The event incidence is due to the modification of physical conditions of the circuit or may be change in the time internal. The physical condition of a circuit involves parametric changes such as temperature of the device, atmospheric pressure, humidity, electromagnetic interference etc. so, when these parametric variations occur the Trojan gets activated as designed by the adversary. The time-based event occurs when the counter present in a circuit reaches a desired count as per the need of an adversary.
- **Externally triggered mechanism:** the externally triggered mechanism involves activation of Trojan present in the circuit when an external input is given to the circuit. The external trigger input may be the user input or it may be because of component output. The external inputs are due to the push-buttons, keyboards, switches, or an external component output connected nearby to the circuit which is having Trojan in it. Thus, hardware Trojan present in the circuit gets activated in the targeted circuit and perform malfunction in the circuit.

D. Effects of trojan

The Trojans are classified based on their effects which are undesirable in nature [8]. Thus, theses undesirable effects lead to a destruction of the target device.

- **Change in the functionality:** the functionality of the design depends on the components present in it, if any unwanted component or a Trojan circuit present within a device it leads to change in the functionality of the targeted device. Thus, the modified specification causes a different functionality instead of intended functionality.
- **Reduction in reliability:** this occurs when there is an intentional modification of device by changing the parameters of the device this causes a reduction in the reliability of the device. The slight or moderate or severe modification of parameters by the adversary results in the variation of power and delay of the device which intern causes the degradation of the device and also leads to aging of the device.
- **Leakage of information:** the Trojan present in the device can leak the information stored in the device. The leakage of information may be caused by the side-channel attacks, or may be because of radio frequency transmission or through various interfaces such as RS-232, JTAG etc. The Trojan even may leak the keys used in the cryptographic algorithms.
- **Denial-of-service (DoS):** the DoS Trojan prohibits the device to perform the certain function or it may stop the device for using the resources. The DoS Trojan may be permanent or temporary in nature. The DoS perform physical damage or it may interrupt the operation of the working device.

E. Location

This describes the presence of Trojans in various locations like processors, memory units, I/O peripherals, clock, power supply, etc. These Trojans are spread widely i.e., it may be of single component or multiple components and they perform independent of each other or they may perform in group.

- In the processing units, Trojans are part of the processing units and they are embedded into the logic gates and thus, these Trojans present in it alters the arrangements of execution of instructions.
- In the memory units, Trojans are part of the memory blocks and their interface units. The presence of Trojan in these blocks will leak the information stored. These Trojan prohibit the user even from assessing the data from the memory.
- In the I/O units, Trojan's attack may be through the peripherals connected to the device. If this Trojans interacts with the external or internal unit of the devices then the Trojan gets a hold on the data communication between the devices thus acquiring the information of the device.
- If the trojan are present in the power supply and the clock grid, then the Trojan cause a this causes a difference in the voltage and the current provided to the chip and the same Trojan present in a clock grid causes an undesirable interrupt in the device.

3 Countermeasures for Hardware Trojan

We summarize the countermeasures against the threats that discussed in the above section. The following explains the summary of the threats from one party to other party in the IC market. In the IC market, to deal with the HT threats we discuss about the defense techniques between each party which ensures trustworthiness in the market [3, 9].

There are three categories of countermeasures against the HT attack: i) hardware trojan detection – it's a process that determine the presence of hardware trojans in the circuit, ii) hardware trojan diagnosis – this is to identify the hardware trojan type and its location in a device, iii) hardware trojan prevention – it is to prevent the hardware trojan insertion during IC development at each stage so that it enhances the efficiency in the trojan detection.

A. Hardware trojan detection

Earlier the trojan detection was highly dependent on the golden IC [10] which is said to be a trojan free IC. If the testing IC differs from that of the golden IC, then that IC is said to be trojan affected, but it is difficult to determine as the complexity of the design increases. Thus, to eliminate the requirement of golden ICs the researchers as proposed advanced techniques to detect the trojan.

- **Trojan detection in pre-silicon stage** – in pre-silicon stage [11], the trojan inserted from the third-party IP cores and the EDA tools are detected using the verification techniques. The trojans are dormant during the functional verification thus it is difficult to determine during verification as they are resistant to traditional verification. Thus, the pre-silicon detection uses set of approaches to detect the trojans. Firstly, it uses the formal verification or assertion-based verification along with the Boolean satisfiability to find the redundant logic present in the circuit. Secondly, it uses the dynamic and static verification techniques formulates the trojan detection as unused circuit identification (UCI) which is considered as suspicious circuits. The UCI algorithm detects the signal pairs and compares the signals with that of golden IC signal and identifies suspicious trojans in the circuit. Recently, to detect the trojans in pre-silicon stage the new approach is proposed i.e., feature analysis-based IP trust vitrification [11]. This method detects the trojan free nets and trojan affected nets without using a golden netlist.

- **Trojan detection in post-silicon stage** – the trojans present in design stage and manufacturing stage of IC design is identified at post-silicon process. In this process the logic test and side-channel analysis approach are used to determine the trojans. Side-channel analysis – this method is used widely for trojan detection, which determines the circuits power consumption, delay, critical path, electromagnetic interference. The SoC designer compare theses parameters and analyze the presence of trojan in the circuit [12]. Logic test – the side-channel analysis approach becomes ineffective due to process variations. Thus, logic test is more effective as they generate the test patterns of the circuit to detect the trojans. as the adversary can insert any number of trojan in the circuit, it is difficult to determine all trojans and the test patterns for each trojan is difficult to generate. Thus, statistical approach for test vector generation has developed.

B. Hardware trojan diagnosis

This approach is used to determine the locations, types and triggers of the trojans in the circuit and to remove the trojans from the circuit. The trojan diagnosis is based on the circuit segmentation and gate level characterization [13]. In this method the large circuits are divided into small sub circuit with lesser number of gates to detect easily and accurately. This also diagnose the trojans by evaluating the leakage power. This process is repeated to obtain the accuracy of the trojan nets.

C. Hardware trojan prevention

To improve the efficiency of the above approaches the prevention are necessary by designing the ICs with self-protection awareness. The techniques for SoC designers to prevent the trojan attacks are done by various methods such as obfuscation, layout-filler, dummy circuit insertion and split manufacturing.

- **Obfuscation** – the obfuscation method is a transformation from one source circuit to a new circuit which has the same functionality as that of the source circuit but the new circuit is less intelligible in some senses [14]. The obfuscation has two strong requirements: i) the obfuscated circuit operates efficiently as same function as that of original circuit. ii) the function of obfuscation circuit will not leak information. Obfuscation techniques are categorized combinatorial logic obfuscation – where there is insertion of additional gates and sequential logic obfuscation – where there is an insertion of extra states.

- **Layout filler** – this technique is proposed to facilitate trojan detection and reduce the insertion of trojans in the empty space of layout. The SoC designer and IP vendors are prohibited from altering the placement and route for trojan circuits. This method prohibits the adversaries in foundries to insert a trojan intentionally or unintentionally in the circuit. In the layout, to avoid the insertion of additional trojans the Built-in self-authentication techniques are used. In this technique the spare space in the layout design is removed and filled with a standard functional cell instead of nonfunctional cells. The standard cells which are inserted in spare space is connected to form a circuit called built-in self-authentication which is independent of original circuit.
- **Split manufacturing** – it's a new technique to protect against the threat from the fabrication at untrusted foundries. This prevents the IP theft and trojan insertion by hiding the part of the design. In this technique the design is partitioned in two or more parts and then fabricated in different foundries so that no one foundry sees the complete design.
- **Run time monitoring** – to reduce the effect of trojan attacks online monitoring should be needed [15]. This monitoring can be used to disable chip when malicious logic is detected or it allow only reliable operation. Chip testing and runtime monitoring are complementary to detect trojans.

4 Machine Learning Models for Trojan Detection

Machine learning algorithm has been used extensively to detect the trojans present in hardware [16]. Thus, to achieve hardware security the machine learning techniques are used in reverse engineering, circuit-feature analysis and side-channel analysis. The Table 1 shows the contribution and innovations of machine learning algorithms applied in trojan detection.

A. ML in Reverse Engineering

Reverse engineering of ICs is the process of analyzing and examining the internal structure of chip design to extract its feature and to obtain the information about the fabrication process. This technique is utilized to reconstruct the original design of the end products and it is destructive process. If any modifications are made in ICs, then using this technique it is easy to achieve the high accuracy by using this technique. This technique is applicable for limited number of IC to obtain the characteristics of golden ICs. By using [17] ML techniques such as SVM (Support Vector Machine) classifier and K-means clustering. These classifiers automatically distinguish between suspicious and expected structures in an IC. The advantage of using ML based classifiers reduces the computation steps as compared to conventional method. These classifiers also generate automated netlist rather manually entering the netlist. The SVM classifier depends on parameters but K-means clustering classifier does not thus, the K-means classifier is easier to train and adjust than SVM. But these classifiers require golden designs which are trojan-free designs.

In the non-destructive reverse engineering, it tries to reverse the netlist of the circuit in terms of state machine by using high-level description of logic circuit. These summaries that it only represents the original function design but not the actual function design. However, the reverse engineering method is costly and time consuming, whether it is of destructive or non-destructive in nature and these methods are valid for simple logic circuits not for complex logic.

B. ML in Circuit Feature Analysis

The trojan activates under special conditions in the circuit as per the requirement of an adversary. These trojans cannot be detected during the traditional testing. Thus, the circuit features like functionality or structural features of the circuit extracts the netlist and analyze the suspicious nets present in the netlist. The two quantitative measures for trojan detection are switching activity and net features. Trojan infected nets from normal nets are differentiated from each net extracted in the features. The SVM [18] or Artificial Neural Networks (ANN) classifier [19] are trained to classify a set of features from an unknown netlist of the circuit. These classifier methods detect the trojan circuit by increasing the True Positive Rate (TPR) – it is defined as prediction accuracy of trojan net, but they have deficiencies in determining True Negative Rate (TNR) – it is defined as prediction accuracy rate of normal net. To avoid this, the Random Forest (RF) classifiers [20] are applied to select the trojan features to detect trojan from these extracted nets.

C. ML in Side-channel Analysis

The trojan detection can be done through side-channel analysis [1] by the affected process variation and noise which relies on signal-to-noise ratio (SNR) and Trojan-to-circuit ratio (TCR). The ANN is used in side-channel analysis, to detect the trojan presented in the circuit by sampling the features extracted [21]. The Back-propagation neural network detection model is used to detect the trojan from the extracted non-linear features by evaluating the power consumption of the circuit. To avoid the inaccuracy in the manual modelling, the Extreme Learning Machine and BPNN (Back-Propagation Neural Network) [22] can be used for the feature extraction [23] and inference as this extract the feature in better way. However, these methodology lack in sampling of features thus, the nets used for training are volatile in nature.

To overcome the instability of the ANN method the SVM is used to train the features and enhance the detection capability and obtain the accuracy. SVM are one of the general classifications of algorithm. It outlines the detection problem and the effects caused by trojan. But this trojan detection methodology has some disadvantage i.e., there is a poor performance in improving the SNR ratio. The other ML methodologies are combination of SVM [17, 18] with other methodologies i.e., SVM+PCA (Principal Component Analysis), DFT (Discrete Fourier Transform) +SVM. The PCA with SVM uses power transmission waveforms to detect the communication type trojans in the circuit. The Discrete Fourier Transform in time domain with SVM are used to covert the waveform data and then trojan detection can be done by SVM. There was another online ML model to detect the trojan. The deep learning was also used to detect trojans.

Thus, these machine learning techniques are very helpful in the trojan detection as they have less computation time even if the circuit is complex.

D. ML in SoC architecture

The vulnerabilities in the architecture level [24] are inserted by an adversary thus, it is necessary to secure the hardware. Hence, hardware security [25, 26] is said to be more necessary part of the research in recent days. As the design for various application are getting to be saturated, the security of these hardware device is necessary. In the architecture level as the design complexity increases the large number of third-party IP cores are used there is a possibility of insertion of trojan in any block of a design.

To enhance the security of the SoC at architecture level [26, 27], the ML can be applied on chip module to secure the device. The analog ANN classifier analyzes the parameters required and classify the device as trojan-free or trojan-infected chip. The attacks by trojans and these trojan can cause the aging through unsupervised strategy. Thus, the confidentiality is prevented by the Runtime trust neural architecture based on adaptive resonance theory. Hence, this method depends on the period of internal clock in an SoC which also eliminate the requirement of golden IC.

5 Results and Discussions

This section discusses about the usage of Machine Learning models in the Hardware Trojan detection. It discusses how frequently the Machine Learning models are used in the detection of Hardware Trojans. The Fig. 4 clearly shows that side-channel analysis uses machine learning models more frequently than other detection technology. The side channel parameters of an ICs are distinguished and it is easy to sample thus side channel analysis uses these distinguished parameters for modelling the classifier.

The circuit feature analysis uses higher machine learning models as compared to reverse engineering. This is because the reverse engineering method involves complexity in the feature extraction and it is time consuming than that of circuit feature analysis thus practicality of these methodology is difficult. The Fig. 4 also shows that among the various machine learning classifier the SVM is the most extensively used classifier however SVM requires a golden ICs as a reference model for training the classifier. Additionally, K means classifier is more in use as compared to other classifiers. However, there are many more algorithms which are widely used as that of SVM thus it is not constrained only for these conditions.

Table 1. Summary of machine learning techniques

Trojan detection	Machine learning based classifiers	Advantage	Disadvantage
Reverse engineering	- SVM [17] - K-means clustering [17]	- Computation time is less - This method simplifies traditional reverse engineering steps from 5 steps to 3 steps - Automatic generation of netlist instead of manually entering	- Golden IC is required as reference model - Classification performance depends on the parameter - These are sensitive to noise and valid only during testing time
Circuit feature analysis	- SVM [18] - ANN [19]	- True positive rate is high - Features are identified automatically and most important features are extracted - Size of the feature vectors are reduced	- True negative rate is very low - Implicit Trojan are not detected - Golden IC is required as a reference model - Execution time depends on the quantitative metric selection
Side-channel analysis	- BPNN [22] - SVM+PCA [17] - SVM+DFT [18]	- Accuracy is high - Relevant features are extracted efficiently with reduced data dimensions - Reduced use of golden ICs	- Poor SNR ratio - Performance of the classifier depends on the parameters, features etc - There is an increase in time consumption

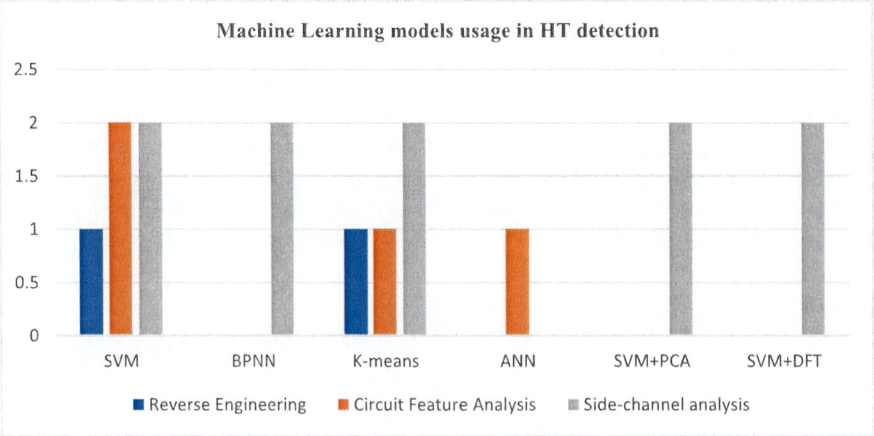

Fig. 4. Usage of machine learning models in HT detection

6 Conclusion

The threats of the hardware trojans are increasing day by day. This has led the researchers to come up with the advanced technologies for preventing and detecting the trojans. In this article, we discussed about the machine-learning based detection methods. We also, discussed about the threats, taxonomy, and counter measures of the hardware trojan. The machine learning technique associated with SoC are also discussed.

The purpose of the paper is to demonstrate work carried out by the researchers so far. It also discusses about the emergence of the machine learning based classifiers in the hardware trojan detection.

References

1. Karri R, Rajendran J, Rosenfeld K, Tehranipoor M (2010) Trustworthy hardware: identifying and classifying hardware Trojans. Computer 43(10):39–46. https://doi.org/10.1109/mc.2010.299
2. Tehranipoor M (2016) New directions in hardware security. In: Proceedings of the 29th international conference on VLSI design, 15th international conference on embedded systems (VLSID), pp 50–52. https://doi.org/10.1109/vlsid.2016.149
3. Bhunia S, Hsiao MS, Banga M, Narasimhan S (2014) Hardware Trojan attacks: threat analysis and countermeasures. Proc IEEE 102(8):1229–1247. https://doi.org/10.1109/jproc.2014.2334493
4. Tehranipoor M, Koushanfar F (2010) A survey of hardware Trojan taxonomy and detection. IEEE Des Test Comput 27(1):10–15. https://doi.org/10.1109/MDT.2010.7
5. Tehranipoor M et al (2011) Trustworthy hardware: Trojan detection and designfor-trust challenges. Computer 44(7):66–74. https://doi.org/10.1109/mc.2010.369
6. Banga M, Hsiao MS (2008) A region-based approach for the identification of hardware Trojans. In: Proceedings of the IEEE international workshop on hardware-oriented security and trust, pp 40–47. https://doi.org/10.1109/hst.2008.4559047
7. Bazzazi A, Taghi MSM, Hemmatyar A (2016) Trojan counteraction in hardware: a survey and new taxonomy. Indian J Sci Technol 9(18):1–9. https://doi.org/10.17485/ijst/2016/v9i18/93764
8. Chakraborty RS, Narasimhan S, Bhunia S (2009) Hardware Trojan: threats and emerging solutions. In: Proceedings of the international high level design validation and test workshop, pp 166–171
9. Sumathi G, Srivani L, Murthy DT, Madhusoodanan K, Murty SS (2018) A review on HT attacks in PLD and ASIC designs with potential defence solutions. IETE Tech Rev 35(1):64–77. https://doi.org/10.1080/02564602.2016.1246385
10. Xue M, Bian R, Liu W, Wang J (2019) Defeating untrustworthy testing parties: a novel hybrid clustering ensemble based golden models-free hardware Trojan detection method. IEEE Access 7:5124–5140
11. Fern N, Cheng K-T (2018) Pre-silicon formal verification of JTAG instruction opcodes for security. In: Proceedings of the IEEE international test conference (ITC), pp 1–9
12. Wang X, Zheng Y, Basak A, Bhunia S (2015) IIPS: infrastructure IP for secure SoC design. IEEE Trans Comput 64(8):2226–2238. https://doi.org/10.1109/tc.2014.2360535
13. Hasegawa K, Yanagisawa M, Togawa N (2017) Hardware Trojans classification for gate-level netlists using multi-layer neural networks. In: Proceedings of the IEEE 23rd international symposium on on-line testing and robust system design (IOLTS), pp 227–232

14. Chakraborty RS, Bhunia S (2011) Security against hardware Trojan attacks using key-based design obfuscation. J Electron Test 27(6):767–785. https://doi.org/10.1007/s10836-011-5255-2
15. Lodhi FK, Hasan SR, Hasan O, Awwadl F (2017) Power profiling of microcontroller's instruction set for runtime hardware Trojans detection without golden circuit models. In: Proceedings of the design, automation & test in Europe conference & exhibition (DATE), pp 294–297. https://doi.org/10.23919/date.2017.7927002
16. Elnaggar R, Chakrabarty K (2018) Machine learning for hardware security: opportunities and risks. J Electron Test 34(2):183–201. https://doi.org/10.1007/s10836-018-5726-9
17. Bao C, Forte D, Srivastava A (2014) On application of one-class SVM to reverse engineering-based hardware Trojan detection. In: Proceedings of the 15th international symposium on quality electronic design, pp 47–54. https://doi.org/10.1109/isqed.2014.6783305
18. Kulkarni A, Pino Y, Mohsenin T (2016) SVM-based real-time hardware Trojan detection for many-core platform. In: Proceedings of the 17th international symposium on quality electronic design (ISQED), pp 362–367. https://doi.org/10.1109/isqed.2016.7479228
19. Madden K, Harkin J, Mcdaid L, Nugent C (2018) Adding security to networks-on-chip using neural networks. In: Proceedings of the IEEE symposium series on computational intelligence (SSCI), pp 1299–1306
20. Hasegawa K, Yanagisawa M, Togawa N (2017) Trojan-feature extraction at gate-level netlists and its application to hardware-Trojan detection using random forest classifier. In: Proceedings of the IEEE international symposium on circuits and systems (ISCAS), pp 1–14
21. Wang S, Dong X, Sun K, Cui Q, Li D, He C (2016) Hardware Trojan detection based on ELM neural network. In: Proceedings of the 1st IEEE international conference on computer communication and the internet (ICCCI), pp 400–403. https://doi.org/10.1109/cci.2016.7778952
22. Li J, Ni L, Chen J, Zhou E (2016) A novel hardware Trojan detection based on BP neural network. In: Proceedings of the 2nd IEEE international conference on computer and communications (ICCC), pp 2790–2794. https://doi.org/10.1109/compcomm.2016.7925206
23. Yang SY, Zhang H (2015) Feature selection and optimization. In: Pattern recognization and intelligent computing, 3rd edn. PHEI, Beijing, pp 27–28. Ch. 2.1, sec. 2
24. Daoud L (2018) Secure network-on-chip architectures for MPSoC: overview and challenges. In: Proceedings of the IEEE 61st international midwest symposium on circuits and systems (MWSCAS), pp 542–543. https://doi.org/10.1109/mwscas.2018.8623831
25. Sustek L (2011) Hardware security module. In: van Tilborg HCA, Jajodia S (eds) Encyclopedia of cryptography and security. Springer, Boston, pp 535–538. https://doi.org/10.1007/978-1-4419-5906-5_509
26. Guha K, Saha D, Chakrabarti A (2015) RTNA: securing SOC architectures from confidentiality attacks at runtime using ART1 neural networks. In: Proceedings of the 19th international symposium on VLSI design and test, pp 1–6. https://doi.org/10.1109/isvdat.2015.7208048
27. Fern N, San I, Koc CK, Cheng K-T-T (2017) Hiding hardware Trojan communication channels in partially specified SoC bus functionality. IEEE Trans Comput Aided Des Integr Circ Syst 36(9):1435–1444. https://doi.org/10.1109/tcad.2016.2638439
28. A detailed description of Trojan. Taxonomy-trust-hub.org

Diagnosis of Covid-19 Patient Using Hyperoptimize Convolutional Neural Network (HCNN)

Maneet Kaur Bohmrah[1](✉) and Harjot Kaur Sohal[2]

[1] Guru Nanak Dev University, Amritsar, India
[2] Guru Nanak Dev University, Gurdaspur, India

Abstract. Due to the Covid-19 virus millions of people lost their lives and livelihood all the over the world, so quick and accurate diagnosis ofcovid-19 patients is the emergent need of today. Deep learning (DL) techniques, nowadays widely gained much popularity as compared to machine learning methods. Furthermore, DL is well-known in the field of image processing, whereas X-ray pictures are frequently used to diagnose Covid-19 patients. In this article, HyperCNN approach introduced through which higher accuracy can be achieved as compared to other traditional Convolutional Neural Network (CNN). Normally, the tuning of hyper parameters is done manually, which are both costly and time consuming in order to identify the optimum model with the highest accuracy. As per the experiment done, the traditional CNN model provides the accuracy 96.6% whereas the proposed model which is tuned using Bayesian optimization achieves accuracy 98.33%. Hence, it is observed that there is an improvement of 2% in the performance of the proposed model as compared to the traditional CNN model. With this experiment, we claim that automatic hyper parameter optimization is a powerful method for improving the accuracy in transfer learning.

Keywords: Hyper parameter · Optimization · Convolutional neural network · Deep learning · Covid-19 diagnosis

1 Introduction

Covid-19 is a deadly virus which completely destroys not only a single country,but the whole world. As this virus spreads exponentially, number of people loses their lives daily [1]. This deadly virus first breakout in the Chinese city of Wuhan in the year 2019 and soon spread around the world. Because it is a contagious disease, it is easily transmitted through the air or by contact with a covid positive person [2]. Most commonly observed symptoms of Covid-19 are cold and fever, swollen throat, dizziness, headache, body ache and Feeling of breathlessness [3].

The World Health Organization (WHO) had declared this disease, which is caused by the severe acute respiratory syndrome coronavirus 2 (SARS-CoV-2), to be a pandemic outbreak [4]. Furthermore, these Covid variations completely impair a human's respiratory system by replicating there, causing dangerous blockages in the lungs and

R. Misra et al. (Eds.): ICIoTCT 2021, LNNS 340, pp. 239–252, 2022.
https://doi.org/10.1007/978-3-030-94507-7_23

weakening the immune system [5, 6]. Covid-19 has become a big issue for practically every country, as it has an economic impact. Because of its contagious nature, almost every government has implemented a strict lockdown [7].

Earlier diagnosis of the Covid-19 is significantly crucial and vital, as prior diagnosis projects the proper treatment on time and helps in controlling the spread of virus too. Numerous ways to diagnose the Covid-19, consisting medical imaging, blood test (CBR) infection rate. Furthermore, X-ray and CT scan are mainly used for this purpose, as it is related to lungs infection which can be identified easily from the X-ray images, if case is worst then CT scan preferred by the doctors [8].

Deep learning subdomain of machine learning, enhance the various applications of artificial intelligence mainly in the field of medical image analysis by achieving the almost human-level accuracies [9]. In this study, the most popular deep learning model, Convolutional Neural Network, was utilized to diagnose Covid-19 using X-ray scans. Hyperparameters of CNN, such as learning rate, number of layers, momentum, and others, are not learned during model training and must be manually assigned before the training phase [10]. On the basis of various surveys, it is found that hyper parameters play a major role in the performance of the deep learning model [11]. As there is no equation formulated for the computation of these hyper parameters, on the other hand hit and trail method proved inappropriate way due to the expensive computation of DL model, which consumes days to train [12].

Furthermore, grid search technique is not suitable as number of computations grows exponentially with the number of hyperparameters. Hence the selection of hyper parameters values introduced as an optimization problem. In this article, Bayesian optimization technique is used to tune the hyper parameters of CNN and improves its accuracy.

The Main contribution of this paper can be mentioned as follows:

- The proposed technique used for the diagnosis of Covid-19 involves HyperCNN model using the Bayesian optimization;
- Number of research papers published for the diagnosis of covid-19 patients till date, we induce the optimization techniques which improves the accuracy of the deep learning model;
- This work introduced the Bayesian optimization for tuning the basic CNN model and it is observed that accuracy improved, hence this paper motivates the researchers to induce the optimization of hyperparameters of deep learning to achieve the more accurate and efficient model.

The rest of the paper is structured as follows as Sect. 2 presents the existing literature work, Sect. 3 describes the terminology used in this paper and provides the description of deep learning CNN and hyper parameters optimization techniques. Sections 4 presents the proposed hyper optimize CNN approach with the extension of this experiment and results were discussed in Sect. 5. Finally, the conclusion was drawn in the Sect. 6.

2 Related Work

In this section, literature work is represented as various articles were published on the diagnosis of Covid-19 after the virus breakout. Hence, few of the articles which were proposed the mechanism for the quick and accurate diagnosis of Covid-19 were mentioned in the Table 1. Moreover, these mentioned research work also used the optimization technique to minimize the CNN architecture automatically hence this work discussed and tried to compare with the proposed mechanism which was introduced in this current article.

Table 1. Review of existing work

Article	Optimization method used	Accuracy
Optimized Convolutional Neural network (OptCoNet) [13]	Grey wolf optimization	97.78%
GSA-DenseNet121-COVID-19 [14]	Gravitational search algorithm	98.38%
CovidXrayNet [15]	Optuna to implement Bayesian optimization	95.82%
[16]	Support vector machine (SVM) and Bayesian optimization	98.97%

3 Theory and Methodology

3.1 Deep Learning and Neural Networks

Deep learning (DL) is the sub domain of Machine learning with biological inspired algorithms. As Neural networks resembles the structure and function of human brain. Neural network mainly consist of three layers namely: Input layer, hidden layer and output layer. Hidden layer is the layers which process the data and extract features which enable the output layer to classify the images or other input data.

When the number of hidden layers increased then it is termed as deep neural network (DNN). With the increase in number of layers, performance and complexity of the network also increased [17]. Deep neural networks handles the large volumes of data due to this feature DNN are more popular as compared other neural networks [18, 19]. *For large scale of data large number of layer network required. Figure 1 depicts the deep learning and Performance of DNN which is directly proportional to the Complexity of DNN.*

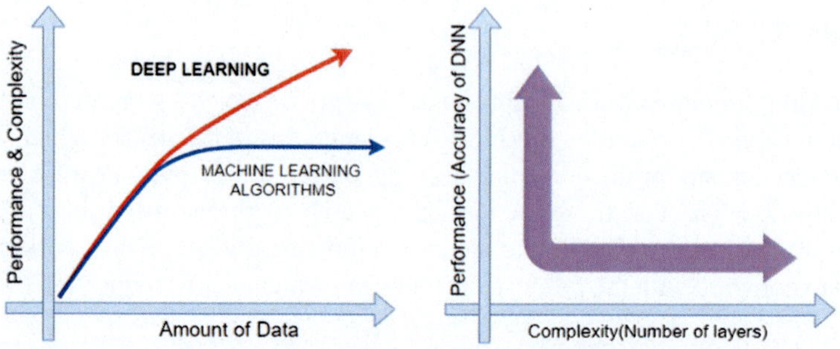

Fig. 1. Characteristics of deep learning

3.2 Convolutional Neural Network (CNN)

CNN plays a key role in the most dynamic and popular fields such as medical imaging, audio processing, stop sign detection, synthetic data generation. This robust deep neural network provides a minimal architecture for undiscovered and learning vital features in image processing [20, 21]. CNN can consist of more than hundreds of layers and each is capable of extracting features of an image. The basic layers are input, hidden and output layer. Hidden layers in CNN also termed as Convolutional block and it is comprises of rectified linear unit and pooling layer. Architecture of Convolutional neural network can be described as:

- **Image input layer** accepts the array of images as input and image should be resize according image input layer size as per the parameters image height, image width and channel. For CNN, normally the input image layer size is (28, 28, 1) where 1 denotes the gray scale images and 3 denotes the RGB which is a color image.
- **Convolution layer** consist of various parameters such as number of filters which is number of neurons that belongs to the same region of input image and another parameter is size of filter. Filters scan the images while training and extracts different kind of image's feature.
- **ReLU layer** also called as Rectified Linear Unit, kind of activation function which speeds up the training of the network by carrying forward only the activated features to further layers.
- **Pooling layer** reduces the number of parameters by removing the redundant features also called down sampling. It can be either max pooling or average pooling.
- **Flatten layer** converts the output taken from the convolutional layers which is in matrix form into a list or vector.
- **Fully connected layer** combines all the features learn from the previous layer to recognize the larger image pattern.
- **Softmax layer** normalizes the output of fully connected layer and also called the output layer, sigmoid function which is also used in place of softmax function. In case of binary classification, sigmoid function is mainly used.

Architecture of CNN depicted in Fig. 2 where one convolutional block comprises of convolutional layer, ReLU and pooling layer is responsible for feature extraction and the output from these layer input to the flatten layer, fully connected layer and softmax which is responsible for classification [22, 23].

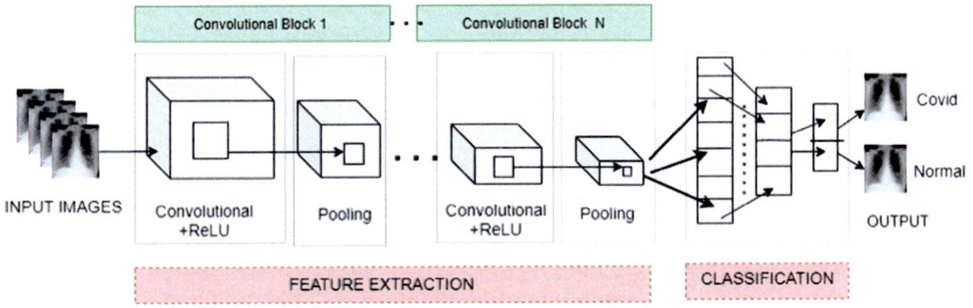

Fig. 2. Architecture of convolutional neural network

3.3 Hyper Parameter Optimization

A deep neural network purely relies on hyper parameters. The choice of values of these hyper parameters directly affects the performance of the network. DNN contains various hyper parameters such as learning rate, batch size, number of epochs, momentum, number of dense layer, number of dense nodes and choice of activation function [24]. Hyper parameters values normally assigned manually by the researchers at the time of training the network but this method is quite time consuming and increase the computational cost also. For instance, number of epochs is the number of passes that the training goes through the complete dataset and one epoch consist of one or more batch. In addition to it, batch size is the number of images in one batch [25]. Traditionally, experts used the hit and trail method for assigning the value for number of epochs, as training of network takes too much time, sometimes it would be hours or even days. Hence, automatic hyper parameters optimization is the necessity of large deep neural networks [26, 27]. Figure 3 represents the process of tuning of hyper parameters in CNN model.

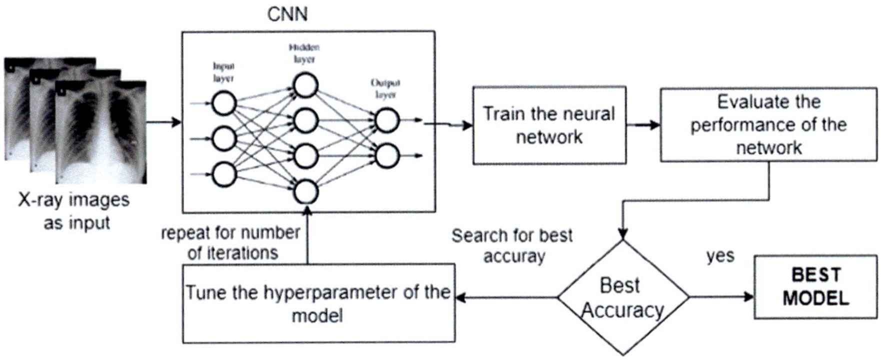

Fig. 3. Hyper parameter optimization for convolutional neural network (CNN)

3.4 Bayesian Optimization

Bayesian Optimization is a powerful technique for global optimization which is widely gained popularity for tuning the hyper parameters of deep learning models for image analysis, audio analysis and various natural language processing. Bayesian optimization is based on iterations with two key parts: a probabilistic surrogate model and assets function to mark which point to compute next. In each round, the surrogate model is outlined to all the observations of the target function. The asset functions are cheap to evaluate and hence thoroughly optimized [28, 29]. Bayesian Optimization technique used to minimize the fitness function which normally takes excessive time to execute. It is normally suited for the optimization of continuous domain having less than 20 dimensions. Bayesian Optimization uses the Gaussian Process which will minimize the expected improvement (EI) and model surrogates' with new probability and repeatedly executed for number of calls to improve the hyper parameters values. Maximum Expected improvement will be used for further calls [30, 31].

4 Proposed Hyper CNN Model

In this paper, we proposed the HyperCNN model which is the integration of CNN and optimization technique. In this article, firstly a simple CNN model created and trained on the Covid-19 dataset. Further, model was tuned with the selected hyper parameters of CNN using Bayesian Optimization and records an improved accuracy. Algorithm 1 describes the process of proposed approach.

Algorithm 1: HyperCNN classification model

Input: Image data set with shape (124,124,1)
Output: Binary classification (covid or normal images with probabilities)
Steps:

1. *Load images in batches of 6 ,images of size (124,124,1)*
2. *Perform the data augmentation technique on image dataset, as its form rotation and scaling of images number of images also increased.*
3. *Select the hyper parameters, for this specific experiment: learning rate, number of dense layer, number of dense node and activation function selected.*
4. *With the default parameter train the Hyper CNN for epochs=3.*
5. *For number of iterations(n)repeat*
   {
   Apply the Gaussian optimization function and train the network with tuned hyper parameter.
   }
6. *Train the model with the hyper parameter values computed by the Bayesian Optimization.*

For tuning the model, four parameters were selected and the search space for all the four parameters were defined initially. For instance, number of layers ranged from 1 to 5 where as number of nodes ranges from 5 to 514 nodes. It can be noted, a huge search space was created. Bayesian Optimization efficiently evaluates the search space for finding the best values for the hyper parameters. Table 2 shows the search space for all the four parameters used in the experiment.

Table 2. Table represents the range of hyper parameters.

Hyper parameter	Abbreviations	Search space
Learning rate	learning_rate	Min = 1e−6, Max = 1e−2
Number of dense layer	num_dense_layers	Min = 1, Max = 5
Number of dense node	num_dense_nodes	Min = 5, Max = 512
Activation function	activation	[Relu, sigmoid]

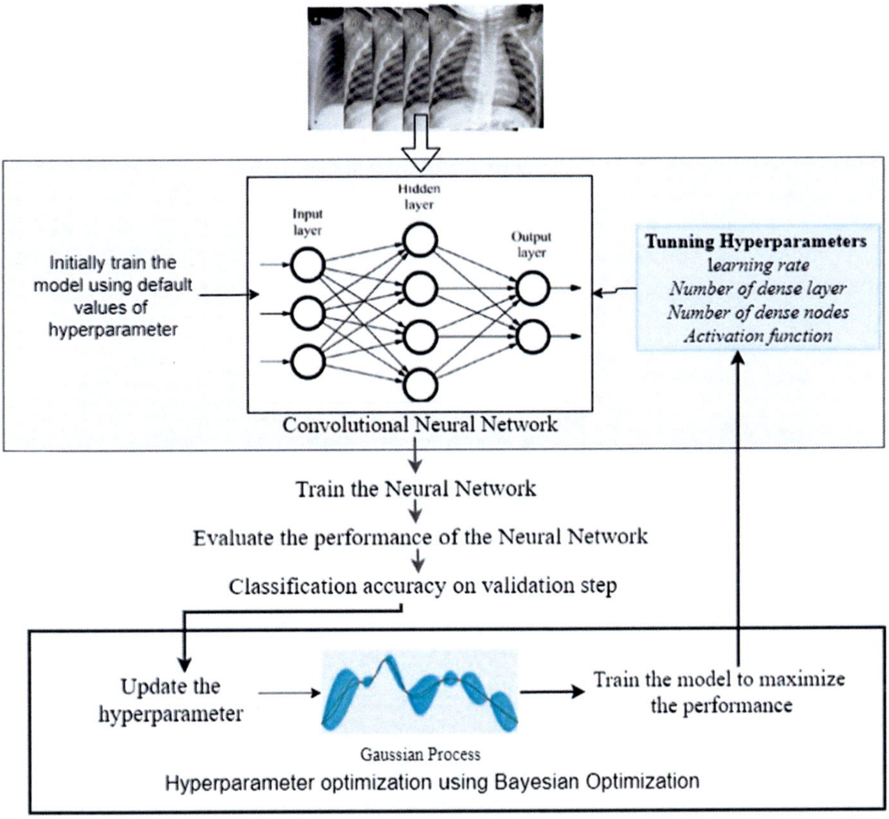

Fig. 4. Proposed HyperCNN model

Figure 4 explained the complete process of HyperCNN. Initially model first trained with the defined default values of parameters. Afterwards, it trained with the hyper parameters tuned by the Bayesian optimization.

5 Experiments and Results

This section explains the dataset used and the analysis of proposed method on the various parameters such as number of iterations and convergence rate. We also discussed a comparative analysis with the traditional CNN model. The complete experiment run on Google Colab with the use of GPU, and Scikit optimize package used for the implementation of Bayesian optimization.

5.1 Dataset Used for the Experiment

The proposed HyperCNN model was trained and tested on a dataset consisting of chest X-ray images from kaggle repository. Dataset contains total 1257 chest X-ray images belongs to two classes: Covid and Normal. Furthermore, we divide our dataset into three sets as follows train, test and validate. Total 288 images belongs to the train dataset which is used to train the model, 60 images belongs to the validate dataset which is used to validate the training conducted on model, and 23 images belongs to the test dataset which is completely unseen images by the model and these images used to test the model, how accurately our trained model recognizes the covid and normal images. Figure 5 shows some sample chest X-rays from the dataset, lungs are surrounded by dark white border which classifies the Covid images and make it differ from the images of normal chest X-rays.

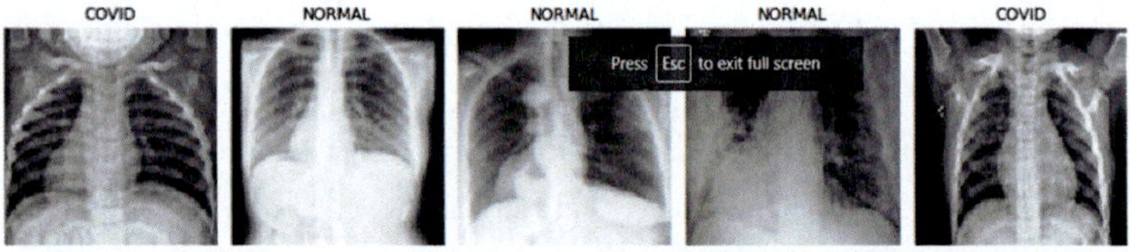

Fig. 5. Sample chest X-ray images from the dataset

5.2 Accuracy Computed by Traditional CNN Model

In this section, we explained the simple CNN model used for the binary classification. Model executed for 25epochs with batch size equals to 6, means one batch contains 6 images. Trained model evaluated on the accuracy and loss function. We have accuracy and loss for both train and validation. As we can see in Fig. 5 loss function falls when models start learning and accuracy goes high. Our model trained from train dataset, and this training validates from the validation dataset, means how much our model learn correctly. More validation accuracy means more accurately our model got trained and hence we focus on improving this validation accuracy. From the experiment, we observed that validation accuracy achieved with the traditional CNN model is 96.67%. Figure 6 describes the relation between training loss and training accuracy. It is clearly observed that as network get trained accuracy increases and loss decreases.

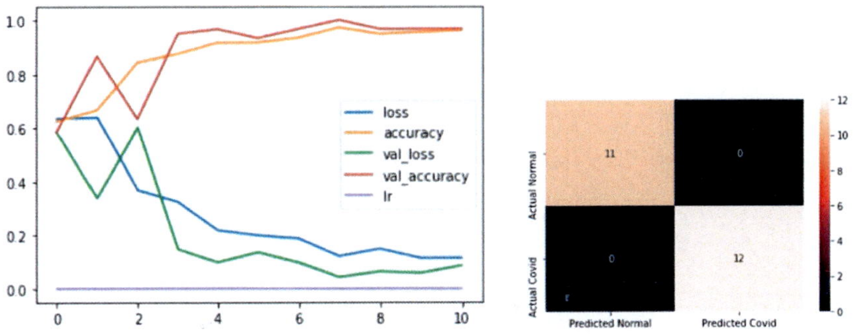

Fig. 6. Performance of traditional CNN model and confusion matrix for traditional CNN on the Covid dataset

5.3 Result Analysis of Proposed Approach HyperCNN

This section describes the result computed by the proposed approach HyperCNN. Bayesian optimization calls for 20, 40, 60 and 80 rounds. It is observed, for 80 rounds we achieved the maximum accuracy. Output with the computed hyper parameter values and accuracyfor80 rounds shown as follows:

```
learning rate: 1.0e-
03num_dense_layers:1
num_dense_nodes:
512activation:sigmoi
d

Epoch3/3
48/48[==============================]-11s227ms/step-
loss:0.3177-accuracy:0.8854-val_loss:0.0884-
val_accuracy:0.9833

Accuracy:98.33%
```

Table 3 depicts the number of iterations, values for each selected hyper parameter computed by the Bayesian Optimization as per the number of calls and the corresponding accuracy.

Table 3. Table shows the computed hyper parameter values and validation accuracies by Bayesian Optimization.

Number of iterations(n)	Computed values of hyper parameters	Accuracy
20	Learning rate = 0.002001701528373809 Number of dense layer = 1 Number of dense nodes = 435 Activation function = 'sigmoid'	96.66%
40	Learning rate = 0.0006573584819425422 Number of dense layer = 1 Number of dense nodes = 358 Activation function = 'sigmoid'	96.66%
60	Learning rate = 0.0008438983592109562 Number of dense layer = 1 Number of dense nodes = 5 Activation function = 'sigmoid'	96.66%
80	Learning rate = 0.0009969613638027 Number of dense layer = 1 Number of dense nodes = 262 Activation function = 'sigmoid'	**98.33%**

Figure 7 describes the values for learning rate, number of dense layer and number of dense nodes. First part shows of plotting is the box plot which shows the maximum and minimum values for all the four hyper parameters computed by HyperCNN whereas second part shows the distribution of these values. On the other hand the Fig. 8 shows the different scatter plots which outline the relation of learning rate, number of dense layer and number of dense nodes with the accuracy. All the following mentioned graphs drawn for the 80 rounds as we achieved maximum accuracy with this. Furthermore, Fig. 9 represents the convergence plot which defines the whole execution of Bayesian Optimization, as our model simultaneously trained and parameters tuned, hence describes the accuracy for corresponding 20 rounds in case of number of calls n = 20. Similarly convergence plot drawn for n = 40, n = 60and n = 80.

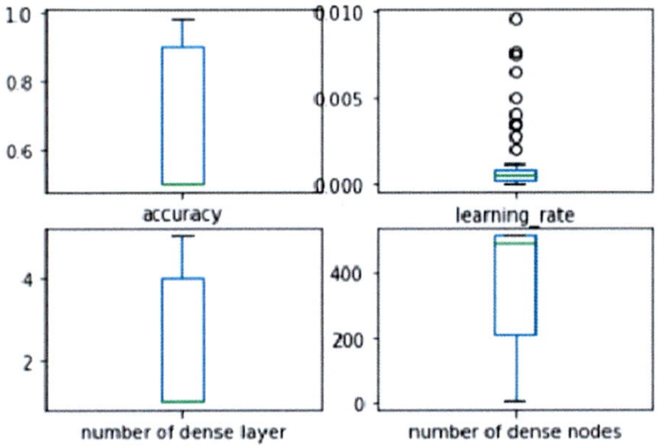

(a)

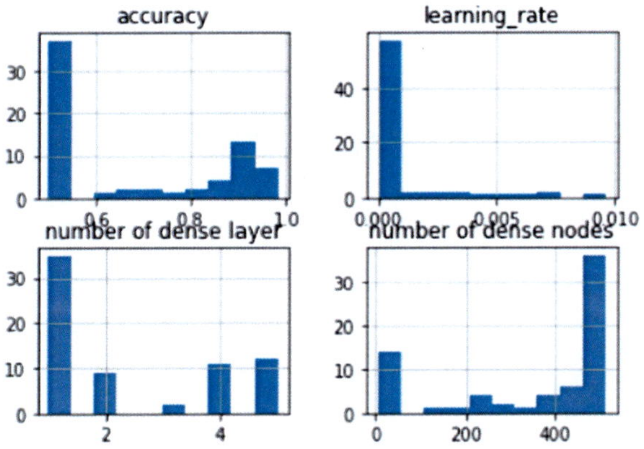

(b)

Fig. 7. Part (a) describes the maximum and minimum value and part (b) describes the distribution of four hyper parameters computed by HyperCNN

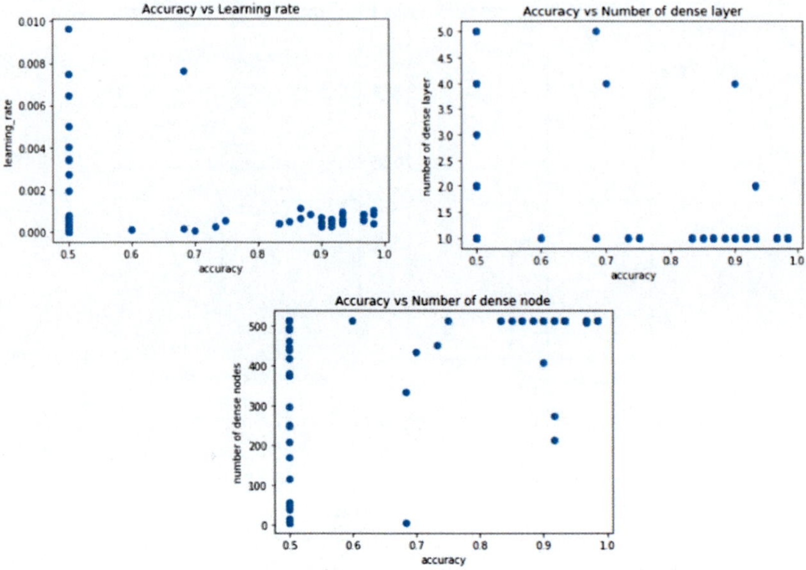

Fig. 8. Scatter plots represents the relation of accuracy with respect to learning rate, number of dense layers and number of dense nodes.

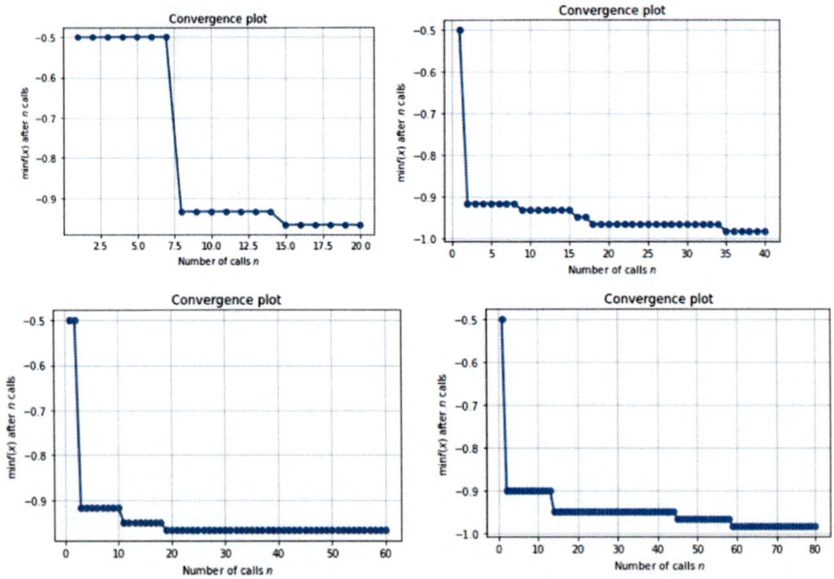

Fig. 9. Convergenceplotfornumberofcalls, n = 20, 40, 60 and 80.

6 Conclusion

In this paper, HyperCNN is proposed which enhanced the validation accuracy of the CNN model. Hyper parameters of CNN were tuned using Bayesian Optimization technique. The dataset of chest X-ray images of Covid-19 patients was collected from kaggle repository and used to train the model for the diagnosis of Covid-19 patients. We execute the HyperCNN for different number of calls and observed for n = 80 results the higher accuracy. It is observed from the experiments that a result from the tuned HyperCNN model was better than the traditional CNN model.

References

1. Chan JF-W et al (2020) A familial cluster of pneumonia associated with the 2019 novel coronavirus indicating person-to-person transmission: a study of a family cluster. The Lancet 395(10223):514–523
2. Peng X, Xu X, Li Y, Cheng L, Zhou X, Ren B (2020) Transmission routes of 2019-nCoV and controls in dental practice. Int J Oral Sci 12(1):1–6
3. Tian S, Hu W, Niu L, Liu H, Xu H, Xiao S-Y (2020) Pulmonary pathology of early-phase 2019 novel coronavirus (COVID-19) pneumonia in two patients with lung cancer. J Thorac Oncol 15(5):700–704
4. Lan TC et al (2020) Structure of the full SARS-CoV-2 RNA genome in infected cells. bioRxiv
5. Razai MS, Doerholt K, Ladhani S, Oakeshott P (2020) Coronavirus disease 2019 (COVID-19): a guide for UK GPs. BMJ 368:m800
6. Remuzzi A, Remuzzi G (2020) COVID-19 and Italy: what next? The Lancet 395(10231):1225–1228
7. Sohrabi C et al (2020) World Health Organization declares global emergency: a review of the 2019 novel coronavirus (COVID-19). Int J Surg 76:71–76
8. Hassanien AE, Mahdy LN, Ezzat KA, Elmousalami HH, Ella HA (2020) Automatic X-ray COVID-19 lung image classification system based on multi-level thresholding and support vector machine. medRxiv
9. Kadry S, Rajinikanth V, Rho S, Raja NSM, Rao VS, Thanaraj KP (2020) Development of a machine-learning system to classify lung CT scan images into normal/COVID-19 class. arXiv preprint arXiv:2004.13122
10. Rubin GD et al (2020) The role of chest imaging in patient management during the COVID-19 pandemic: a multinational consensus statement from the Fleischner Society. Radiology 296(1):172–180
11. Mazurowski MA, Buda M, Saha A, Bashir MR (2019) Deep learning in radiology: an overview of the concepts and a survey of the state of the art with focus on MRI. J Magn Reson Imaging 49(4):939–954
12. Khorrami P, Paine T, Huang T (2015) Do deep neural networks learn facial action units when doing expression recognition? In: IEEE international conference on computer vision, Santiago, Chile, pp 19–27
13. Goel T, Murugan R, Mirjalili S, Chakrabartty DK (2020) OptCoNet: an optimized convolutional neural network for an automatic diagnosis of COVID-19. Appl Intell 51(3):1351–1366. https://doi.org/10.1007/s10489-020-01904-z
14. Ezzat D, Hassanien AE, Ella HA (2020) An optimized deep learning architecture for the diagnosis of COVID-19 disease based on gravitational search optimization. Appl Soft Comput 98:106742

15. Monshi MM, Poon J, Chung V, Monshi FM (2021) CovidXrayNet: optimizing data augmentation and CNN hyperparameters for improved COVID-19 detection from CXR. Comput Biol Med 133:104375
16. Nour M, Cömert Z, Polat K (2020) A novel medical diagnosis model for COVID-19 infection detection based on deep features and Bayesian optimization. Appl Soft Comput 97:106580
17. Zoph B, Vasudevan V, Shlens J, Le QV Learning transferable architectures for scalable image recognition. arXiv:1707.07012
18. Zhong Z, Yan J, Wei W, Shao J, Liu C-L (2018) Practical block-wise neural network architecture generation. In: Conference on computer vision and pattern recognition, Salt Lake City, Utah, USA. arXiv preprint arXiv:1708.05552
19. Razavian AS, Azizpour H, Sullivan J, Carlsson S (2014) CNN features off-the-shelf: an astounding baseline for recognition. In: Conference on computer vision and pattern recognition, Columbus, Ohio, USA, pp 806–813
20. Nilsback M-E, Zisserman A (2008) Automated flower classification over a large number of classes. In: Conference on computer vision, graphics image processing, Madurai, India, pp 722–729
21. Bergstra J, Yamins D, Cox D (2013) Making a science of model search: hyperparameter optimization in hundreds of dimensions for vision architectures. In: International conference on machine learning, Atlanta, USA, pp 115–123
22. Dan Claudiu C, Meier U, Gambardella LM, Schmidhuber J (2011) Convolutional neural network committees for handwritten character classification. In: International conference on document analysis and recognition, pp 1135–1139
23. McDonnell MD, Vladusich T (2018) Enhanced image classification with a fast-learning shallow convolutional neural network. In: International joint conference on neural networks. IEEE, pp 1–7
24. Domhan T, Springenberg JT, Hutter F (2015) Speeding up automatic hyperparameter optimization of deep neural networks by extrapolation of learning curves. In: Yang Q, Wooldridge M (eds) Proceedings of the 25th international joint conference on artificial intelligence (IJCAI 2015), pp 3460–3468
25. Bergstra J, Bardenet R, Bengio Y, Kégl B (2011) Algorithms for hyper-parameter optimization. In: Shawe-Taylor J, Zemel R, Bartlett P, Pereira F, Weinberger K (eds) Proceedings of the 25th international conference on advances in neural information processing systems (NeurIPS 2011), pp 2546–2554
26. Bergstra J, Bengio Y (2012) Random search for hyper-parameter optimization. J Mach Learn Res 13:281–305
27. Garrido-Merchán E, Hernández-Lobato D (2017) Dealing with integer-valued variables in Bayesian optimization with Gaussian processes. arXiv:1706.03673v2 [stats.ML]
28. Gelbart M, Snoek J, Adams R (2014) Bayesian optimization with unknown constraints. In: Zhang N, Tian J (eds) Proceedings of the 30th conference on uncertainty in artificial intelligence (UAI 2014). AUAI Press
29. Wang J, Xu J, Wang X (2048) Combination of hyperband and Bayesian optimization for hyperparameter optimization in deep learning. arXiv:1801.01596v1 [cs.CV]
30. Snoek J, Larochelle H, Adams RP (2012) Practical Bayesian optimization of machine learning algorithms. In: Advances in neural information processing systems, pp 2951–2959
31. Zatarain Cabada R, Rodriguez Rangel H, Barron Estrada ML, Cardenas Lopez HM (2019) Hyperparameter optimization in CNN for learning-centered emotion recognition for intelligent tutoring systems. Soft Comput 24(10):7593–7602. https://doi.org/10.1007/s00500-019-04387-4

Comparison of Resampling Methods on Mobile Apps User Behavior

Isuru Dharmasena[1], Mike Domaratzki[2], and Saman Muthukumarana[1(✉)]

[1] Department of Statistics, University of Manitoba, Winnipeg, MB R3T2N2, Canada
dharmrai@myumanitoba.ca, Saman.Muthukumarana@umanitoba.ca
[2] Department of Computer Science, University of Manitoba, Winnipeg, MB R3T2N2, Canada
Michael.Domaratzki@cs.umanitoba.ca

Abstract. Mobile applications have become a vital part in modern businesses where products and services are offered in real-time. As many people have adopted to mobile apps, it is not uncommon that some of the applications are used for a few times and then abandoned. This "churning" effect on mobile apps has become a wide topic of interest among businesses to understand the factors affecting the user abandonment. This includes predicting and identifying the abandoning users beforehand to actively engage users to have more active and loyal app users. With datasets for churning, there is often a class imbalance problem where the retained user group is the minority class. We study and assess several over-sampling methods and under-sampling methods combined with several classification methods to improve the prediction ability and model performance of mobile app user retention using data available from a local mobile app developing company. The results indicate that combining under-sampling and over-sampling techniques improve overall model performance and right pick of re-sampling techniques are critical for better predictive results.

Keywords: Class imbalance · Classification · Re-sampling · Customer retention · ROC

1 Introduction

Generally, most of the learning algorithms and learning systems assume that the data used to learn are balanced with equal instances in each class of the response variable. However, in the real world it is not always true. The number of instances in one class might be more abundant than the others which tend to obstruct the performance of classifiers obtained through Machine Learning (ML) algorithms.

A dataset is said to be imbalanced if the number of instances of each class of the response variable are not approximately equal. The imbalance can be of two types, between-class imbalance and within-class imbalance. Between-class

R. Misra et al. (Eds.): ICIoTCT 2021, LNNS 340, pp. 253–271, 2022.
https://doi.org/10.1007/978-3-030-94507-7_24

imbalance is where some classes have more instances than others [5]. Within-class imbalance on the other hand refers to scenarios where subsets of one class have fewer instances than other subsets of same class [42]. Furthermore, the classes with more instances are identified as majority classes or groups while the classes with lesser instances are identified as minority classes or groups in imbalanced datasets.

Imbalance in the proportion of 100 to 1 is frequent in fraud detection and imbalance proportion up to 100,000 to 1 has been observed in other applications [39]. Imbalanced data is found in many real world classification problems such as detection of oil spills in satellite radar images [27], telecommunication customer management [15], text classification [13,30,31] and caller user profiling [16].

Challenges in Learning from Imbalanced Data Sets

Learning from an imbalanced dataset is often considered as a challenging task. By considering a hypothetical simplified two class problem given in Fig. 1, we can see that Fig. 1a) shows a imbalance between minority instances (+) and majority instances (–) and Fig. 1b) shows a balanced data set with distinguishable clusters. In Fig. 1a) class overlapping also makes it difficult for some classifiers such as k-Nearest Neighbors (k-NN). Minority instances surrounded by majority instances often lead to miss-classification later which will diminish the overall predictive ability of the learning algorithm.

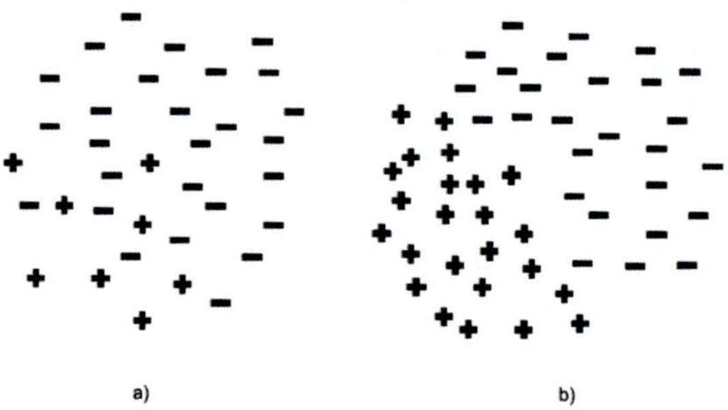

Fig. 1. a) Imbalanced case with sparse positive cases b) Balanced data set with well separated clusters

Overlapping minority and majority instances makes it difficult to separate the classes since the lack of instances in one class make the border-line weak. Previous works have shown many ways to overcome these problems. One aspect is to balance the data set by means of over-sampling techniques. Another aspect is cleaning the imbalanced data sets by tidying up noisy majority/minority instances such that border-line of the classes becomes stronger and more distinguishable.

This imbalance issue has been addressed primarily in two ways in previous studies [2–4,14]. One is to use unique cost values to train instances [10,35].

The other method is to re-sample the original data, either by under-sampling the majority class and/or over-sampling the minority class [25,28,30,32]. Under-sampling can be considered as cleaning the dataset where the classifier is mislead by noisy majority instances. Some over-sampling techniques play a role of making the classifiers stronger with synthetically generated minority instances where the classifier tends to fail in certain situations such as near the borderline.

Mobile Applications and Customer Retention

In this era of advanced communication technologies, mobile applications (apps) have become primary tools in people's personal and professional lives. Mobile Apps facilitate multiple applications including but not limited to communication, social media, education, entertainment, medical, utilities and travel. Mobile Apps are not only important for the App users but also it plays a crucial role in many modern businesses. Any company providing their services through mobile Apps are interested in new user acquisition as well as retention of existing customers. The customers who continue to use the mobile App over a given period can be considered as retained users whereas this is the opposite of churned users.

Some might argue that increased number of downloads of a particular mobile App indicates a better metric on how well that App is retained among customers but it is not always true. One person might download an App and abandon using after one day or may keep the mobile App without using it for a long time. So far there is no clear line indicating a mobile App user to be classified as retained or churned. Generally it is defined by the App provider considering facts such as the business model and the nature of the mobile App. Independent of the nature of business, the majority of the mobile Apps face a churn rate approximately around 70% after 90 days [36]. This retention rate of 30% indicates that only 30 out of 100 mobile App users tend to return to their mobile Apps or "loyal" to the App. Furthermore, it has been shown that 22% use an App only once after downloading it [24].

The dataset we use in this study contains in-app feature usage of a fitness mobile app from a local app developing company. When considering our predictive problem using this dataset on mobile App user retention, it is an imbalanced problem with 3075 (80.3%) instances of users that left the mobile App (majority class) and only 755 (19.7%) instances of users who retained (minority class). This problem is a binary classification problem where the App users who have left are re-coded as 0 and the retained customers are re-coded as 1. The retention of the App users were defined by the company based on the weekly usage of the mobile App by each customer. In our experiment, we use several re-sampling techniques, over-sampling and under-sampling techniques together to improve the overall performance of several classifiers. Then the improvement of classifiers are tested and verified using permutation based hypothesis tests.

2 Re-sampling Methods

Re-sampling can be done with either over-sampling or under-sampling depending on the problem. Previous studies have implemented several over-sampling techniques to create new instances for the minority group until the dataset becomes balanced.

Over-Sampling Methods

Random Over-Sampling (ROS) can be considered as the most widely used over-sampling techniques before other innovative methods have been discovered. Minority class samples are randomly selected and replicated to achieve the balanced dataset [17]. One issue with random over-sampling is that this method duplicates already existing data which would not necessarily benefit the classification algorithm since duplicates would not give new information on how to classify new observations. Moreover, this often tends to increase the likelihood of overfitting and there it may reduce the classifier performance while increasing the computational time.

Synthetic Minority Over-sampling TEchnique (SMOTE) was introduced as an innovative method of producing "synthetic" instances of minority class without duplicating already existing minority class instances [7]. SMOTE finds the k-nearest neighbors of each instance in minority class and the identified nearest neighbors are used to create new instances by randomly choosing a point on the line connecting the instance with the nearest neighbor repeatedly. This avoids any overfitting issue as in ROS.

Borderline-SMOTE (BLSMOTE) is a derivative of SMOTE which over-samples minority instances near the borderline between classes [21]. This method focuses on borderline class instances and using Borderline-SMOTE1 and Borderline-SMOTE2, only the borderline instances are over-sampled given their importance in classification rather than instances which are away from the borderline. Unlike SMOTE, this method tries to over-sample and "strengthen" the borderline minority examples by first identifying the borderline minority examples and adding synthetically generated instances to the original training dataset.

ADAptive SYNthetic sampling (ADASYN) is a successor of previous synthetic data generation techniques such as SMOTE, SMOTEBoost [6] and DataBoostIM [19]. The most important objective of introducing this method is to reduce the bias and adaptively learning from the given data [20].

Majority Weight Minority Over-sampling TEchnique (MWMOTE) is another improvement done over existing over-sampling techniques so that the minority instances that are harder to learn will be isolated and assigned weights according to their Euclidean distances to the nearest minority class instances. With this method all the instances generated synthetically will fall within the minority cluster [1].

Under-Sampling Methods

In under-sampling, we downsize the actual dataset such that the response variable categories become at most 10:1.

Random Under-sampling (RUS) involves removal of random instances from the majority class with or without replacement. This is considered the earliest under-sampling techniques used. This may increase the variance of the classifier, hence potentially may discard useful and important instances from the original dataset [17].

Edited Nearest Neighborhood Rule (ENN) algorithm removes instances from the a class that are missclassified by their k nearest neighbors [40,43]. Often Euclidean metrics are used to classify an candidate instance to the class of its nearest neighbor in the measurement space.

Neighborhood Cleaning Rule (NCL) uses Wilson's Edited Nearest Neighbor (ENN) rule [43] to remove majority instances and get rid of any instance whose class differ from the class of at least two of its three nearest neighbors [29].

Tomek Links (TL) can be defined as follows; given two instances E_i and E_j in different classes, and $d(E_i, E_j)$ is the distance between E_i and E_j. A (E_i, E_j) pair is called a TL if there is no instance E_l such that $d(E_i, E_l) < d(E_i, E_j)$ or $d(E_j, E_l) < d(E_i, E_j)$. If two instances create a TL, then either one or both instances are on the borderline. This method can be used to under-sample and clean the borderline majority instances [40].

One-Sided Selection (OSS) is another under-sampling technique which uses Tomek Links (TL) to detect and remove borderline majority instances and applies the Condensed Nearest Neighborhood Rule (CNN) [23] to further under-sample by removing majority instances that are away from the borderline [28].

Under-Sampling Based on Clustering (SBC) uses k clusters to randomly sample majority instances from each cluster based on the imbalance percentage within those clusters [44].

3 Classifiers

We compare effects of combined re-sampling methods over three classification algorithms. Although supervised learning algorithms depend on dependent and independent class probabilities of the training instances, each classifying algorithm uses different approaches to find a solution to the problem. In this section, we would introduce basic concepts of the three classifiers to understand the way each re-sampling method affect each classifier.

Logistic Regression

The logistic regression model is widely implemented in binary classification problems [18] as it provides predictions in the form of 0 or 1 values. It usually fits data with maximum likelihood method and models the probabilities of the data class and response variables as a linear function. Assuming this linearity in the function with only one explanatory variable (x), the logistic function can be written as follows:

$$\hat{y} = \frac{1}{1 + \exp^{-(\beta_0 + \beta_1 x)}} \tag{1}$$

where β_1 is the coefficient of the explanatory variable x and β_0 is the intercept.

We note that the fact that logistic regression requires far less computational resources compared to some classifiers like support vector machines (SVM) can be considered as a benefit. Furthermore, the linear function of the logistic model provides the significance of each response variable towards the outcome of the response variable [18].

Naïve Bayes

Naïve Bayes classifier is a probabilistic classifier that assumes the independence between the response variables. The probability of each response variable given the class is learned by the algorithm and then it predicts the class of a given sample based on sum of probabilities of the explanatory variables. The Naïve Bayes classifier which assigns the class label $\hat{y} = C_k$ for k number of possible outcomes of classes C_k for a problem with n explanatory variables $x = (x_1, x_2, ..., x_n)$ is as follows:

$$\hat{y} = \underset{k \in 1,0}{\operatorname{argmin}}\, p(C_k) \prod_{i=1}^{n} p(x_i | C_k) \tag{2}$$

The assumption on the independence of explanatory variables makes it efficient in high dimensional data sets and it also requires less computational power though the algorithm seems straightforward [18].

Support Vector Machine (SVM)

Hyperplanes are used to classify datasets with a high dimensional feature space in support vector machine (SVM). In order to find the optimal hyperplane that would maximize the distance between margins, a SVM uses kernels such as radial basis function (RBF) to calculate distance between high dimensional data points. The Support vectors are the optimal marginal data points that anchor the hyperplane. The function to predict the class of a new sample with weights, where the weights of the hyperplane that provide the maximum margin which is trained on the train set is as follows:

$$\hat{y} = w.u + b = (\sum_{i=1}^{l} a_i y_i x_i).u + b \tag{3}$$

where x_i are the input features with set of weights w whose linear combination predicts y_i s for l instances with bias value b and a_i slack variables that are introduced in the maximization problem.

Table 1. Confusion matrix for binary classification problem

Predicted		
Actual	Positive	Negative
Positive	True Positive (TP)	False Negative (FN)
Negative	False Positive (FP)	True Negative (TN)

Support Vector Regression (SVR)

Support Vector Regression (SVR) [12] is the regression version of SVM which is often used in high dimensional regression problems. Interestingly, SVR maintains all the properties from SVM while attempting to find a match between some vector and the position in the curve found by SVR which is not acting as a decision boundary. Support vectors participate in finding the best match between data instances and the actual function that is represented by them. When the distance between support vectors and regression curve is maximized, it become more close to the actual curve. Like SVM, SVR can also use kernels in order to regress non-linear functions. In our problem, we use a variation of SVR, which is nu-SVR which the number of support vectors limited. For the purpose of this study, we use `kernlab` R package [26] and the `ksvm()` function with only changing the type to `nu-svr` and kernel to `rbfbot` that will yield SVR model by using nu-SVR and radial-basis kernal. The rest of parameters are kept in their default values as assigned in the R package itself. This implementation is more computational intensive that previous two classifiers in our study.

4 Evaluation Metrics of Classifiers

Performance of classifiers have been primarily assessed using tools such as precision, recall and accuracy to reflect the effect of imbalanced data [32,39]. More information about the actual and predicted classes of a given binary classifier can be obtained using a confusion matrix in Table 1 shown below:

Here in Table 1, represents a confusion matrix of a binary classification problem having **positive** (1) and **negative** (0) class values. It is possible to extract number of widely used performance metrics like precision, recall, accuracy, F1 score from a confusion matrix like in Table 1. The methods of computing previously mentioned performance metrics are as follows:

$$\text{Precision} = \frac{TP}{TP + FP} \tag{4}$$

$$\text{Recall} = \frac{TP}{TP + FN} \tag{5}$$

$$\text{Accuracy} = \frac{TP + TN}{TP + TN + FP + FN} \tag{6}$$

$$\text{F1} - \text{score} = 2 \times \frac{Precision.Recall}{Precision + Recall} \tag{7}$$

The above performance matrices that uses values from both classes in a confusion matrix like Table 1 would be sensitive to class skewness and might mislead especially in an imbalance situation. For example, when we use accuracy or error rate (1-accuracy) it is a disadvantage in an imbalance problem since it considers both classification errors (either positive or negative) to be equally important. To address this issue, it would be better to consider metrics that consider classes independently as follows:

$$\text{False negative rate} = FN_{rate} = \frac{FN}{TP + FN} \tag{8}$$

$$\text{False positive rate} = FP_{rate} = \frac{FP}{FP + TN} \tag{9}$$

$$\text{True negative rate} = TN_{rate} = \frac{TN}{FP + TN} \tag{10}$$

$$\text{True postive rate} = TP_{rate} = \frac{TP}{TP + FN} \tag{11}$$

These performance measures are independent from class probabilities and costs. Furthermore, Receiver Operating Characteristic (ROC) curve [33,38] can be used to analyze the relationship between FN rate and FP rate (or TN rate and TP rate). It characterizes the performance of a binary classifier across all trade offs between the sensitivity of the classifier (TP_{rate}) and the false alarm (FP_{rate}). ROC analysis also allows the comparison of multiple classification functions simultaneously. Furthermore, area under curve (AUC) of ROC curve represents the expected model performance in a single scalar and is equivalent to the Wilcoxon rank test and other statistical measures of evaluating classification and ranking models [22]. F1 score can also be considered as a sound measurement for classification problems since it encircles the trade-off between precision and recall and reflects how well a classifier is in a single measurement [37].

k-Fold Cross Validation

In k-fold cross validation, a given dataset D is partitioned into k equal and mutually exclusive partitions (folds) $D_1, D_2, ..., D_k$. Then each partition is used to test the model which is trained on the remainder partitions combined together as the training set. Hence, k-fold cross validation make sure that the candidate model is trained on many possible combinations of data to obtain a better estimate on the model metrics preventing possible overfitting. Although k-fold cross validation is computationally intensive, reduced bias in the results and decrease in variance of the estimate with the increasing of number of folds (k) can be considered as the key advantages. Typically the value of k set to 5 or 10.

In our study, we obtain multiple ROC curves for every re-sampling strategy. In order to compare those ROC curves, there are few methods proposed in literature. One is to fit a parametric model and test the equality of the parameters [11,34].

A redefined non-parametric test was introduced by DeLong [9] to compare the AUC for paired and unpaired data. Furthermore, Venkatraman [41] have developed a complete non-parametric test to compare two ROC curves when the data are paired and continuous. This test is also capable of distinguishing two ROC curves cross each other but have equal AUC s using non-parametric hypothesis tests. We use 10-fold cross validation so to obtain a ROC curve that represent every scenario is challenging. In literature, several methods are discussed for multi-reader multi-case (MRMC) ROC studies in medical imaging systems [8]. Here, we will discuss about the methods prevailing on averaging ROC curves and then try choose a method of averaging ROC curves for our analysis.

5 Methods

Mobile App user dataset consist of 27 explanatory variables and a response variable with only 19.7% of instances for App users that are retained at the end of the time period. The explanatory variables are the in-app features that were available for the users to interact over 12 weeks. We combine over-sampling and under-sampling methods to treat the imbalance percentage of the response variable and use logistic regression, Naïve Bayes and Support Vector Machines to classify retention of the mobile App users. We use four levels $(35\%, 40\%, 45\%, 50\%)$ of over-sampling percentages with respect to the whole dataset and then under-sample any imbalance scenario using under-sampling techniques so that the final response variable is balanced. For an example, 50% represents balancing the response variable using the over-sampling technique while 40% means that we over-sample minority instances such that the new imbalance percentage of the response variable is 40% and then use under-sampling techniques to downsize the majority instances so that the final training dataset is a balanced dataset.

To compare the performances of the classifiers, we use F1 score, area under curve (AUC) and ROC curves. Furthermore permutation-based hypothesis testings are used to distinguish similar ROC curves.

6 Results

Consider 10-fold cross validation for the logistic regression model using the original training dataset as in Fig. 2a. To obtain an average ROC curves we have few options.

- Average by calculating mean TPR and FPR values from folds
- Average sensitivity (S_e) at each specificity (S_p)
- Average specificity at each sensitivity
- Average $\frac{S_e+S_p}{2}$ at each fixed $\frac{S_e-S_p}{2}$.

The first method is simply taking the mean value of each TPR and FPR value to get a mean ROC curve as shown in Fig. 2b. We can generalize last three options by following algorithm [8]:

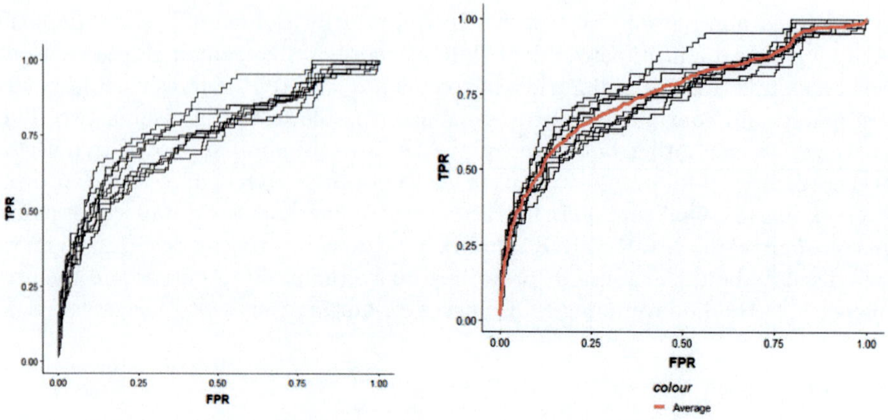

(a) ROC curves for 10-fold cross validation (b) Mean ROC curve

Fig. 2. ROC curves for 10-fold cross validation with mean ROC curve

- Rotate the axes (FPR, TPR) in ROC space counter-clockwise for an angle θ to the (u, v) space:

$$\begin{cases} u = FPRcos\theta + TPRsin\theta \\ v = -FPRsin\theta + TPRcos\theta \end{cases}$$

- Average ROC curves in (u, v) space by averaging v for each u
- Rotate the averaged curve in (u, v) space back to ROC space:

$$\begin{cases} FPF = ucos\theta - vsin\theta \\ TPF = -usin\theta + vcos\theta \end{cases}$$

The parameter θ influences the direction along which the ROC curves are averaged. With this algorithm, the method of averaging sensitivity (S_e) at each specificity (S_p) is when $\theta = 0$ (Fig. 3). Similarly, averaging specificity at each sensitivity corresponds to $\theta = \frac{\pi}{2}$ (Fig. 4) while the last method is when $\theta = \frac{\pi}{4}$ (Fig. 5).

Furthermore, we will retain all data of every ROC curve in 10-fold cross validation and try to obtain a smoothed ROC curve that represent an average ROC curve as given in Fig. 6. When we compare the corresponding resultant average curves for each can be compared together as shown in Fig. 7. According to this plot the average ROC curves seems to be similar but when we consider area under curves for each averaging method (Table 2), we can observe that ROC curve obtained by averaging specificity at each sensitivity yields the least AUC value.

For the purpose of assessing the performance of classifiers with re-sampling strategies, we use the method of retaining all 10-fold prediction data in order to obtain an average ROC curve for a given model.

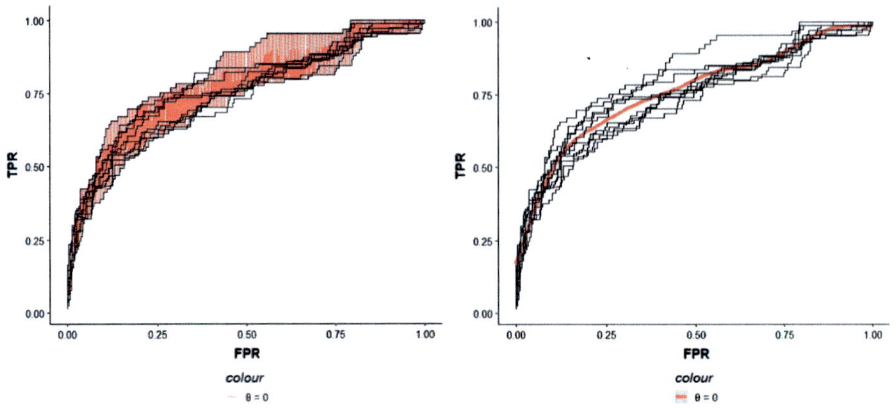

(a) Average ROC curve for $\theta = 0$

(b) Smoothed average ROC curve for $\theta = 0$

Fig. 3. Average ROC curve and respective smoothed ROC curve by averaging sensitivity at each specificity ($\theta = 0$)

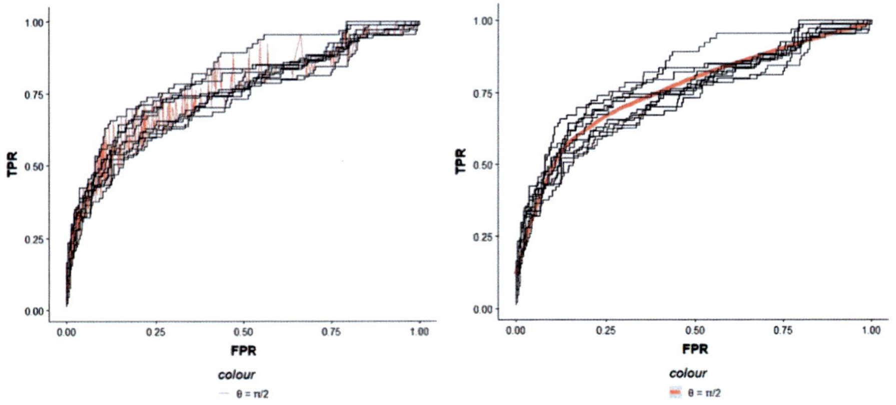

(a) Average ROC curve for $\theta = \frac{\pi}{2}$

(b) Smoothed average ROC curve for $\theta = \frac{\pi}{2}$

Fig. 4. Average ROC curve and respective smoothed ROC curve by averaging specificity at each sensitivity ($\theta = \frac{\pi}{2}$)

Table 2. Average AUC from each averaging method

Method	Average AUC
All	0.7612
Mean	0.7563
$\theta = 0$	0.7615
$\theta = \pi/2$	0.7166
$\theta = \pi/4$	0.7570

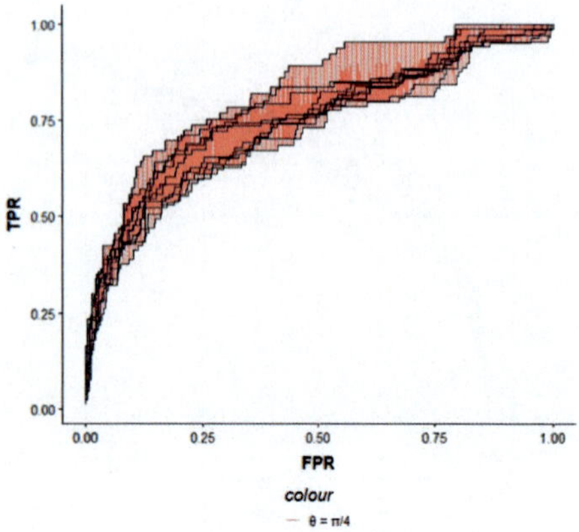

(a) Average ROC curve for $\theta = \frac{\pi}{4}$

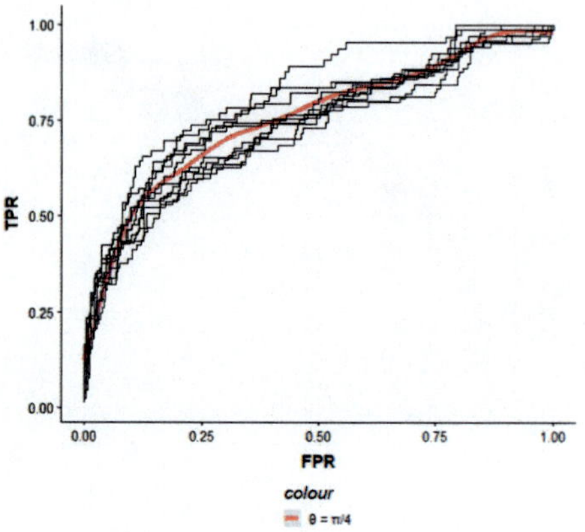

(b) Smoothed average ROC curve for $\theta = \frac{\pi}{4}$

Fig. 5. Average ROC curve and respective smoothed ROC curve by averaging $\frac{S_e+S_p}{2}$ at each fixed $\frac{S_e-S_p}{2}$ $(\theta = \frac{\pi}{4})$

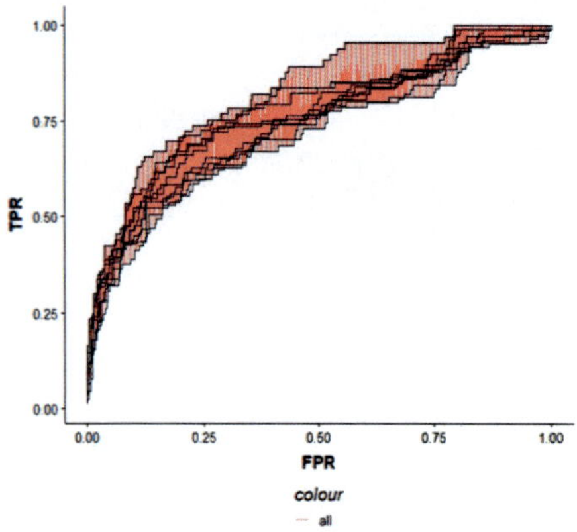

(a) Average ROC curve by combining all 10-fold data

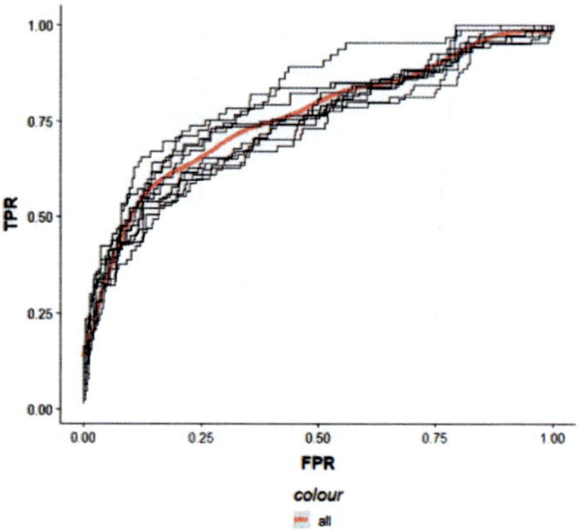

(b) Smoothed average ROC curve by combining all 10-fold data

Fig. 6. Average ROC curve and respective smoothed ROC curve by smoothing all 10-fold ROC data combined

Mobile App user data was used to train on re-sampled training dataset and classify on 958 testing instances. Considering all combinations of over-sampling percentages, over-sampling techniques and under-sampling techniques Table 3 shows the top performing combinations of each classifier according to F1 scores. Table 4 shows the most under performing combinations for each classifier while Table 5 shows classifier performance obtained from training dataset without using any re-sampling technique. As seen in Table 3, the top performing classifier according to F1 score is logistic regression model obtained using a train dataset obtained by over-sampling minority instances using SMOTE until the percentage of minority group is 40% and then cleaning the majority group using

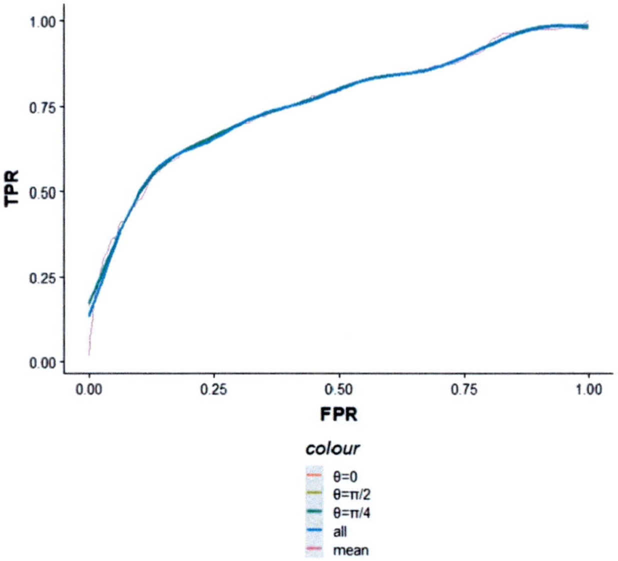

Fig. 7. Average ROC curves from all ROC curve averaging methods

Table 3. Highest F1 score for each classifier by re-sampling methods with 10-fold cross validation

Percentage	Over-sampling method	Under-sampling method	Model	F1 score	sd_F1	AUC	sd_AUC
40	SMOTE	ENN	Logistic	0.5270	0.0254	0.7633	0.0261
50	MWMOTE	OSS	SVR	0.4952	0.0402	0.7312	0.0372
40	MWMOTE	ENN	NaiveBayes	0.4621	0.0298	0.7353	0.0317

Table 4. Least F1 score for each classifier by re-sampling methods with 10 fold cross validation

Percentage	Over-sampling method	Under-sampling method	Model	F1 score	sd_F1	AUC	sd_AUC
35	No oversampling	ENN	Logistic	0.3846	0.0477	0.7535	0.0324
45	BLSMOTE_1	RUS	NaiveBayes	0.2501	0.1021	0.6021	0.0919
50	No oversampling	TL	SVR	0.0443	0.0209	0.7197	0.0379

ENN. The second best classifier is SVR which used a train dataset that was balanced using MWMOTE and then clean the majority group using OSS.

When considering F1 scores and area under curve (AUC) values of the classifiers that were trained on re-sampling techniques as shown in Fig. 8 and Fig. 9, it can be seen that both of the measurements follow a similar changes over the percentage of over-sampling. Overall for most of the resampling techniques, logistic regression classifier's F1 score and AUC values tend to improve with the

Table 5. Classifier performance by F1 score without using any re-sampling strategies with 10-fold cross validation

Model	F1 score	sd_F1	AUC	sd_AUC
NaiveBayes	0.4079	0.0498	0.7374	0.0418
Logistic	0.3846	0.0477	0.7535	0.0324
SVR	0.0500	0.0218	0.7168	0.0363

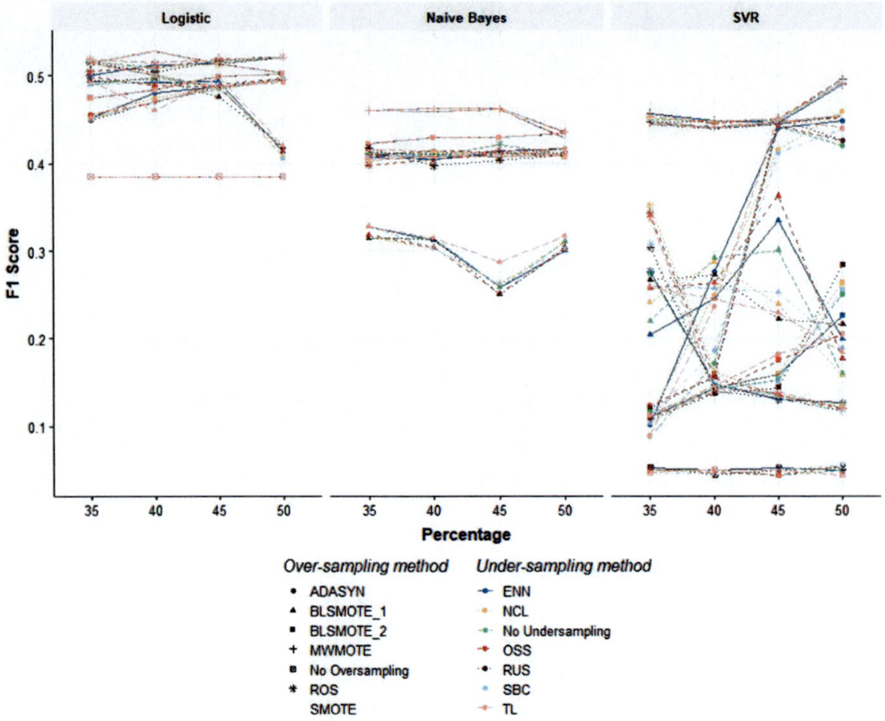

Fig. 8. F1-scores of classifiers with percentage of over-sampling

percentage of over-sampling of minority instances although those values drastically diminish when Borderline-SMOTE 1 is used to balance the dataset. This is true for SVR classifier but for Naïve Bayes classifier, over-sampling with Borderline-SMOTE 1 has improved the performance comparatively.

To compare and differentiate the similarities between ROC curves with similar AUC values, we use De Long's non-parametric hypothesis testing for the best three models with highest F1 scores from each classifier as shown in Table 3. The respective p-values are given in Table 6.

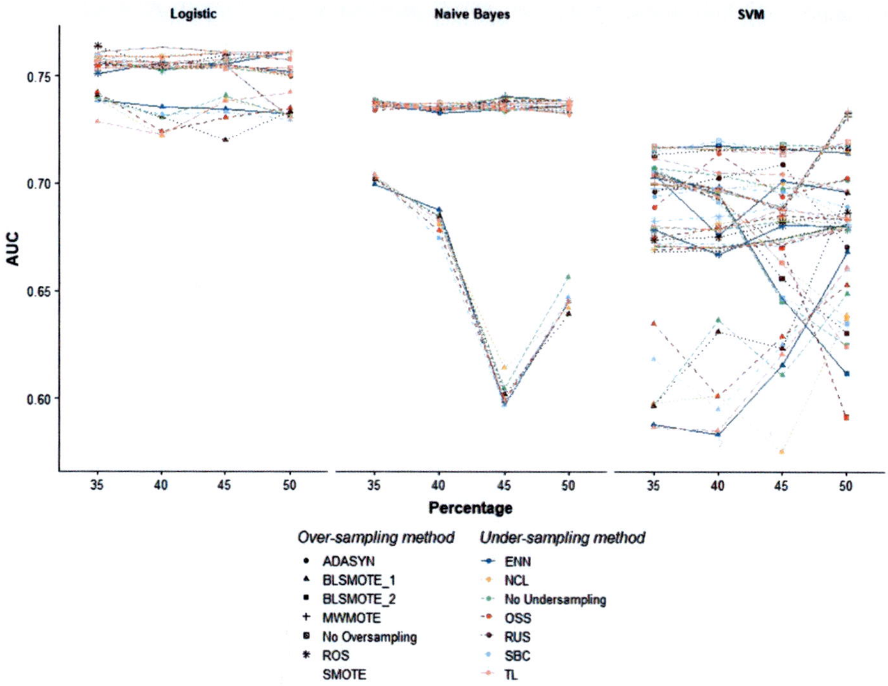

Fig. 9. Area under curves of classifiers with percentage of over-sampling

Table 6. p-values by De Long's non-parametric hypothesis tests to compare Area Under Curves of Table 3

vs	Logit	NaiveBayes	SVR
Logit		0.00006*	0.00145*
NaiveBayes	0.00006*		0.31605
SVR	0.00145*	0.31605	

7 Conclusion

In our study we analyzed the behavior of several over and under-sampling methods in learning imbalanced datasets. Our results show that for over-sampling methods, SMOTE, MWMOTE and Borderline-SMOTE 1 with under-sampling techniques such as Edited Nearest Neighbors and One-Sided Selection together yields better metrics on classification performance. Moreover random over-sampling and random under-sampling which is generally considered to be under-performing, resulted in competitive results compared to more complex re-sampling techniques. Another interesting observation is that some re-sampling strategies degraded the classifier performance than improving. For an example, comparing Table 5 and Table 4, we can see that Näive Bayes and

SVR performance reduced with certain re-sampling combinations. Eliminating critical majority instances with under-sampling techniques could be a reason for those results. The classifier performance improved significantly using over-sampling techniques such as SMOTE and MWMOTE than other over-sampling techniques. Overall all the classifiers' performances were improved using over-sampling combined with under-sampling showing the importance of supporting the classification algorithms by means of re-sampling. With the availability of many re-sampling strategies, the recommendation is to consider the technique of over-sampling as well as the amount of minority samples to be over-sampled prior to under-sampling the majority group.

De Long's hypothesis testing can be used as a tool to distinguish similar ROC curves for situations as similar to this study. ROC curves for the best näive Bayes and SVR models are similar and with the hypothesis testing it yields a p-value of 0.31605 failing to reject null hypothesis at 95% confidence level (Table 6) implying that the two ROC curves are similar, i.e. the performance of the two models over each cut-off point is approximately similar while the ROC curve for the best logistic regression model differ from the näive Bayes model and the SVR model according to the hypothesis test p-values which are both less that 0.05.

Another aspect of our experiment was to under-sample majority instances and then over-sample the minority instances. This approach raised several concerns including the dataset obtained by under-sampling techniques becomes smaller and hence the classifier performance degraded.

It should be noted that the results gained for the mobile App user data might be different for any other scenario with different imbalance ratio. The results would be different for each imbalance level of the train dataset. For any given scenario, the strategy of using over-sampling and under-sampling techniques discussed in this study can be applied to find how the model performance changes over the re-sampling techniques. This will assist on picking the suitable re-sampling strategy to treat the imbalance problem. Identifying the optimal imbalance ratio for an initial train set would be addressed in future research.

References

1. Barua S, Islam MM, Yao X, Murase K (2014) MWMOTE - majority weighted minority oversampling technique for imbalanced data set learning. IEEE Trans Knowl Data Eng 26(2):405–425. https://doi.org/10.1109/TKDE.2012.232
2. Batuwita R, Palade V (2010) Efficient resampling methods for training support vector machines with imbalanced datasets. In: The 2010 international joint conference on neural networks (IJCNN). IEEE, pp 1–8
3. Burnaev E, Erofeev P, Papanov A (2015) Influence of resampling on accuracy of imbalanced classification. In: Eighth international conference on machine vision (ICMV 2015), vol 9875. International Society for Optics and Photonics, p 987521
4. Cateni S, Colla V, Vannucci M (2014) A method for resampling imbalanced datasets in binary classification tasks for real-world problems. Neurocomputing 135:32–41

5. Chawla N, Japkowicz N, Kołcz A (2004) Editorial: special issue on learning from imbalanced data sets. SIGKDD Explor 6:1–6. https://doi.org/10.1145/1007730.1007733
6. Chawla NV (2005) Data mining for imbalanced datasets: an overview. Springer, Boston, pp 853–867. https://doi.org/10.1007/0-387-25465-X_40
7. Chawla NV, Bowyer KW, Hall LO, Kegelmeyer WP (2002) SMOTE: synthetic minority over-sampling technique. Technical report
8. Chen W, Samuelson FW (2014) The average receiver operating characteristic curve in multireader multicase imaging studies. Br J Radiol 87(1040):20140016
9. DeLong E, DeLong D, Clarke-Pearson D (1988) Comparing the areas under two or more correlated receiver operating characteristic curves: a nonparametric approach. Biometrics 44(3):837–845. https://doi.org/10.2307/2531595
10. Domingos P (1999) MetaCost: a general method for making classifiers cost-sensitive. In: Proceedings of the fifth ACM SIGKDD international conference on knowledge discovery and data mining, KDD 1999. Association for Computing Machinery, New York, pp 155–164. https://doi.org/10.1145/312129.312220
11. Dorfman DD, Alf E (1969) Maximum-likelihood estimation of parameters of signal-detection theory and determination of confidence intervals-rating-method data. J Math Psychol **6**(3):487–496. https://doi.org/10.1016/0022-2496(69)90019-4. http://www.sciencedirect.com/science/article/pii/0022249669900194
12. Drucker H, Burges CJ, Kaufman L, Smola AJ, Vapnik V (1997) Support vector regression machines. In: Advances in neural information processing systems, pp 155–161
13. Dumais S, Platt J, Heckerman D, Sahami M (1998) Inductive learning algorithms and representations for text categorization. In: Proceedings of the seventh international conference on information and knowledge management, CIKM 1998. Association for Computing Machinery, New York, pp 148–155. https://doi.org/10.1145/288627.288651
14. Estabrooks A, Jo T, Japkowicz N (2004) A multiple resampling method for learning from imbalanced data sets. Comput Intell 20(1):18–36
15. Ezawa K, Singh M, Norton SW (1996) Learning goal oriented Bayesian networks for telecommunications risk management. In: Proceedings of the 13th international conference on machine learning. Morgan Kaufmann, pp 139–147
16. Fawcett T, Provost F (1996) Combining data mining and machine learning for effective user profiling. In: Proceedings of the second international conference on knowledge discovery and data mining, KDD 1996. AAAI Press, pp 8–13
17. Fernández A, García S, Herrera F, Chawla NV (2018) SMOTE for learning from imbalanced data: progress and challenges, marking the 15-year anniversary. https://doi.org/10.1613/jair.1.11192
18. Friedman J, Tibshirani R, Hastie T (2009) The elements of statistical learning. Springer, New York
19. Guo H, Viktor HL (2004) Learning from imbalanced data sets with boosting and data generation. Technical report 1. https://doi.org/10.1145/1007730.1007736
20. He H, Bai Y, Garcia, EA, Li S (2008) ADASYN: adaptive synthetic sampling approach for imbalanced learning. In: 2008 IEEE international joint conference on neural networks (IEEE world congress on computational intelligence). IEEE, pp 1322–1328. https://doi.org/10.1109/IJCNN.2008.4633969. http://ieeexplore.ieee.org/document/4633969/
21. Han H, Wang WY, Mao BH (2005) Borderline-SMOTE: a new over-sampling method in imbalanced data sets learning. Technical report

22. Hand DJ (1997) Construction and assessment of classification rules. Wiley, Chichester
23. Hart P (1968) The condensed nearest neighbor rule (corresp.). IEEE Trans Inf Theory 14(3):515–516
24. Hoch D (2014) App retention improves - apps used only once declines to 20%. http://info.localytics.com/blog/app-retention-improves
25. Japkowicz N (2000) The class imbalance problem: significance and strategies. In: Proceedings of the 2000 international conference on artificial intelligence (ICAI), pp 111–117
26. Karatzoglou A, Smola A, Hornik K, Karatzoglou MA (2019) Package 'kernlab'. CRAN R Project
27. Kubat M, Holte RC, Matwin S (1998) Machine learning for the detection of oil spills in satellite radar images. Mach Learn 30(2):195–215. https://doi.org/10.1023/A:1007452223027
28. Kubat M, Matwin S (1997) Addressing the curse of imbalanced training sets: one-sided selection. In: ICML
29. Laurikkala J (2001) Improving identification of difficult small classes by balancing class distribution. In: Quaglini S, Barahona P, Andreassen S (eds) Artificial Intelligence in Medicine. AIME 2001. LNCS, vol 2101. Springer, Heidelberg, pp 63–66. https://doi.org/10.1007/3-540-48229-6_9
30. Lewis DD, Catlett J (1994) Heterogeneous uncertainty sampling for supervised learning. In: Cohen WW, Hirsh H (eds) Machine learning proceedings 1994. Morgan Kaufmann, San Francisco, pp. 148–156. https://doi.org/10.1016/B978-1-55860-335-6.50026-X. http://www.sciencedirect.com/science/article/pii/B978155860335650026X
31. Lewis DD, Ringuette M (1994) A comparison of two learning algorithms for text categorization
32. Ling CX, Li C (1998) Data mining for direct marketing: problems and solutions. In: KDD
33. Maloof MA (2003) Learning when data sets are imbalanced and when costs are unequal and unknown. In: ICML-2003 workshop on learning from imbalanced data sets II, vol 2, pp 1–2
34. Metz CE, Wang PL, Kronman HB (1984) A new approach for testing the significance of differences between ROC curves measured from correlated data. In: Deconinck F (eds) Information Processing in Medical Imaging. Springer, Dordrecht, pp 432–445. https://doi.org/10.1007/978-94-009-6045-9_25
35. Pazzani M, Merz C, Murphy P, Ali K, Hume T, Brunk C (1994) Reducing misclassification costs. In: Cohen WW, Hirsh H (eds) Machine learning proceedings 1994. Morgan Kaufmann, San Francisco, pp 217–225. https://doi.org/10.1016/B978-1-55860-335-6.50034-9. http://www.sciencedirect.com/science/article/pii/B9781558603356500349
36. Perro J (2018) Mobile apps: what's a good retention rate? http://info.localytics.com/blog/mobile-apps-whats-a-good-retention-rate
37. Powers D (2011) Evaluation: from precision, recall and F-measure to ROC, informedness, markedness & correlation. J Mach Learn Technol 2:2229–3981. https://doi.org/10.9735/2229-3981
38. Provost, F, Fawcett T (1997) Analysis and visualization of classifier performance: comparison under imprecise class and cost distributions. In: Proceedings of the third international conference on knowledge discovery and data mining, KDD 1997. AAAI Press, pp 43–48

39. Provost F, Fawcett T (2001) Robust classification for imprecise environments. Mach Learn 42(3):203–231. https://doi.org/10.1023/A:1007601015854
40. Tomek I (1976) An experiment with the nearest-neighbor rule. IEEE Trans Syst Man Cybernet SMC–6(6):448–452
41. Venkatraman ES, Begg CB (1996) A distribution-free procedure for comparing receiver operating characteristic curves from a paired experiment. Biometrika 83(4):835–848. https://doi.org/10.1093/biomet/83.4.835
42. Weiss GM (2004) Mining with rarity: a unifying framework. SIGKDD Explor Newsl 6(1):7–19. https://doi.org/10.1145/1007730.1007734
43. Wilson DL (1972) Asymptotic properties of nearest neighbor rules using edited data. IEEE Trans Syst Man Cybern 2(3):408–421. https://doi.org/10.1109/TSMC.1972.4309137
44. Yen SJ, Lee YS (2009) Cluster-based under-sampling approaches for imbalanced data distributions. Expert Syst Appl 36(3, Part 1):5718–5727. https://doi.org/10.1016/j.eswa.2008.06.108. http://www.sciencedirect.com/science/article/pii/S0957417408003527

Author Index

R. Misra et al. (Eds.): ICIoTCT 2021, LNNS 340, pp. 273–274, 2022.
https://doi.org/10.1007/978-3-030-94507-7